TEACHER RESOURCES

Living Systems

Full Option Science System
Developed at the Lawrence Hall of Science, University of California, Berkeley
Published and Distributed by Delta Education

FOSS Lawrence Hall of Science Team
Larry Malone and Linda De Lucchi, FOSS Project Codirectors and Lead Developers
Kathy Long, FOSS Assessment Director; David Lippman, Program Manager; Carol Sevilla, Publications Design Coordinator; Susan Stanley, Illustrator; John Quick, Photographer
FOSS Curriculum Developers: Brian Campbell, Teri Lawson, Alan Gould, Susan Kaschner Jagoda, Ann Moriarty, Jessica Penchos, Kimi Hosoume, Virginia Reid, Joanna Snyder, Erica Beck Spencer, Joanna Totino, Diana Velez, Natalie Yakushiji
Susan Ketchner, Technology Project Manager
FOSS Technology Team: Dan Bluestein, Christopher Cianciarulo, Matthew Jacoby, Kate Jordan, Frank Kusiak, Nicole Medina, Jonathan Segal, Dave Stapley, Shan Tsai

Delta Education Team
Bonnie A. Piotrowski, Editorial Director, Elementary Science
Project Team: Mathew Bacon, Jennifer Apt, Sandra Burke, Tom Guetling, Joann Hoy, Angela Miccinello

Thank you to all FOSS Grades K-5 Trial Teachers
Heather Ballard, Wilson Elementary, Coppell, TX; Mirith Ballestas De Barroso, Treasure Forest Elementary, Houston, TX; Terra L. Barton, Harry McKillop Elementary, Melissa, TX; Rhonda Bernard, Frances E. Norton Elementary, Allen, TX; Theresa Bissonnette, East Millbrook Magnet Middle School, Raleigh, NC; Peter Blackstone, Hall Elementary School, Portland, ME; Tiffani Brisco, Seven Hills Elementary, Newark, TX; Darrow Brown, Lake Myra Elementary School, Wendell, NC; Heather Callaghan, Olive Chapel Elementary, Apex, NC; Katie Cannon, Las Colinas Elementary, Irving, TX; Elaine M. Cansler, Brassfield Road Elementary School, Raleigh, NC; Kristy Cash, Wilson Elementary, Coppell, TX; Monica Coles, Swift Creek Elementary School, Raleigh, NC; Shirley Conner, Ocean Avenue Elementary School, Portland, ME; Sally Connolly, Cape Elizabeth Middle School, Cape Elizabeth, ME; Melissa Cook-Airhart, Harry McKillop Elementary, Melissa, TX; Melissa Costa, Olive Chapel Elementary, Apex, NC; Hillary P. Croissant, Harry McKillop Elementary, Melissa, TX; Rene Custeau, Hall Elementary School, Portland, ME; Nancy Davis, Martha and Josh Morriss Mathematics and Engineering Elementary School, Texarkana, TX; Nancy Deveneau, Wilson Elementary, Coppell, TX; Karen Diaz, Las Colinas Elementary, Irving, TX; Marlana Dumas, Las Colinas Elementary, Irving, TX; Mary Evans, R.E. Good Elementary School, Carrollton, TX; Jacquelyn Farley, Moss Haven Elementary, Dallas, TX; Corinna Ferrier, Oak Forest Elementary, Humble, TX; Allison Fike, Wilson Elementary, Coppell, TX; Barbara Fugitt, Martha and Josh Morriss Mathematics and Engineering Elementary School, Texarkana, TX; Colleen Garvey, Farmington Woods Elementary, Cary, NC; Judy Geller, Bentley Elementary School, Oakland, CA; Erin Gibson, Las Colinas Elementary, Irving, TX; Kelli Gobel, Melissa Ridge Intermediate School, Melissa, TX; Dollie Green, Melissa Ridge Intermediate School, Melissa, TX; Brenda Lee Harrigan, Bentley Elementary School, Oakland, CA; Cori Harris, Samuel Beck Elementary, Trophy Club, TX; Kim Hayes, Martha and Josh Morriss Mathematics and Engineering Elementary School, Texarkana, TX; Staci Lynn Hester, Lacy Elementary School, Raleigh, NC; Amanda Hill, Las Colinas Elementary, Irving, TX; Margaret Hillman, Ocean Avenue Elementary School, Portland, ME; Cindy Holder, Oak Forest Elementary, Humble, TX; Sarah Huber, Hodge Road Elementary, Knightdale, NC; Susan Jacobs, Granger Elementary, Keller, TX; Carol Kellum, Wallace Elementary, Dallas, TX; Jennifer A. Kelly, Hall Elementary School, Portland, ME; Brittani Kern, Fox Road Elementary, Raleigh, NC; Jodi Lay, Lufkin Road Middle School, Apex, NC; Melissa Lourenco, Lake Myra Elementary School, Wendell, NC; Ana Martinez, RISD Academy, Dallas, TX; Shaheen Mavani, Las Colinas Elementary, Irving, TX; Mary Linley McClendon, Math Science Technology Magnet School, Richardson, TX; Adam McKay, Davis Drive Elementary, Cary, NC; Leslie Meadows, Lake Myra Elementary School, Wendell, NC; Anne Mechler, J. Erik Jonsson Community School, Dallas, TX; Anne Miller, J. Erik Jonsson Community School, Dallas, TX; Shirley Diann Miller, The Rice School, Houston, TX; Keri Minier, Las Colinas Elementary, Irving, TX; Stephanie Renee Nance, T.H. Rogers Elementary, Houston, TX; Cynthia Nilsen, Peaks Island School, Peaks Island, ME; Elizabeth Noble, Las Colinas Elementary, Irving, TX; Courtney Noonan, Shadow Oaks Elementary School, Houston, TX; Sarah Peden, Aversboro Elementary School, Garner, NC; Carrie Prince, School at St. George Place, Houston, TX; Marlaina Pritchard, Melissa Ridge Intermediate School, Melissa, TX; Alice Pujol, J. Erik Jonsson Community School, Dallas, TX; Claire Ramsbotham, Cape Elizabeth Middle School, Cape Elizabeth, ME; Paul Rendon, Bentley Elementary, Oakland, CA; Janette Ridley, W.H. Wilson Elementary School, Coppell, TX; Kristina (Crickett) Roberts, W.H. Wilson Elementary School, Coppell, TX; Heather Rogers, Wendell Creative Arts & Science Magnet Elementary School, Wendell, NC; Alissa Royal, Melissa Ridge Intermediate School, Melissa, TX; Megan Runion, Olive Chapel Elementary, Apex, NC; Christy Scheef, J. Erik Jonsson Community School, Dallas, TX; Samrawit Shawl, T.H. Rogers School, Houston, TX; Nicole Spivey, Lake Myra Elementary School, Wendell, NC; Ashley Stephenson, J. Erik Jonsson Community School, Dallas, TX; Jolanta Stern, Browning Elementary School, Houston, TX; Gale Stimson, Bentley Elementary, Oakland, CA; Ted Stoeckley, Hall Middle School, Larkspur, CA; Cathryn Sutton, Wilson Elementary, Coppell, TX; Camille Swander, Ocean Avenue Elementary School, Portland, ME; Brandi Swann, Westlawn Elementary School, Texarkana, TX; Robin Taylor, Arapaho Classical Magnet, Richardson, TX; Michael C. Thomas, Forest Lane Academy, Dallas, TX; Jomarga Thompkins, Lockhart Elementary, Houston, TX; Mary Timar, Madera Elementary, Lake Forest, CA; Helena Tongkeamha, White Rock Elementary, Dallas, TX; Linda Trampe, J. Erik Jonsson Community School, Dallas, TX; Charity VanHorn, Fred A. Olds Elementary, Raleigh, NC; Kathleen VanKeuren, Lufkin Road Middle School, Apex, NC; Valerie Vassar, Hall Elementary School, Portland, ME; Megan Veron, Westwood Elementary School, Houston, TX; Mary Margaret Waters, Frances E. Norton Elementary, Allen, TX; Stephanie Robledo Watson, Ridgecrest Elementary School, Houston, TX; Lisa Webb, Madisonville Intermediate, Madisonville, TX; Matt Whaley, Cape Elizabeth Middle School, Cape Elizabeth, ME; Nancy White, Canyon Creek Elementary, Austin, TX; Barbara Yurick, Oak Forest Elementary, Humble, TX; Linda Zittel, Mira Vista Elementary, Richmond, CA

Photo Credits: © Andrey Armyagov/Shutterstock (cover); © Zach Smith; © Monkey Business Images/Shutterstock; © John Quick; © Christian Musat/Shutterstock

Published and Distributed by Delta Education, a member of the School Specialty Family
The FOSS program was developed in part with the support of the National Science Foundation grant nos. MDR-8751727 and MDR-9150097. However, any opinions, findings, conclusions, statements, and recommendations expressed herein are those of the authors and do not necessarily reflect the views of NSF. FOSSmap was developed in collaboration between the BEAR Center at UC Berkeley and FOSS at the Lawrence Hall of Science.

Copyright © 2016 by The Regents of the University of California

Standards cited herein from NGSS Lead States. 2013. *Next Generation Science Standards: For States, By States.* Washington, DC: The National Academies Press. Next Generation Science Standards is a registered trademark of Achieve. Neither Achieve nor the lead states and partners that developed the Next Generation Science Standards was involved in the production of, and does not endorse, this product.

All rights reserved. Any part of this work (other than duplication masters) may not be reproduced or transmitted in any form or by any means, electronic or mechanical, including photocopying and recording, or by an information storage or retrieval system without prior written permission. For permission please write to: FOSS Project, Lawrence Hall of Science, University of California, Berkeley, CA 94720 or foss@berkeley.edu.

Living Systems — Teacher Toolkit, 1487685
Teacher Resources, 1487599
978-1-62571-352-0
Printing 5 – 9/2017
Patterson Printing, Benton Harbor, MI

TEACHER RESOURCES

TABLE OF CONTENTS

FOSS Program Goals

Science Notebooks in Grades 3–5

Science-Centered Language Development

FOSS and Common Core ELA — Grade 5

FOSS and Common Core Math — Grade 5

Taking FOSS Outdoors

Science Notebook Masters

Teacher Masters

Assessment Masters

This document, *Teacher Resources*, is one of three parts of the *FOSS Teacher Toolkit* for this module. The chapters in *Teacher Resources* are all available as PDFs on FOSSweb.

The other parts of the module *Teacher Toolkit* are the *Investigations Guide* and a copy of the *FOSS Science Resources* student book containing original readings for this module.

The spiral-bound *Investigations Guide* contains these chapters.

- Overview
- Framework and NGSS
- Materials
- Technology
- Investigations
- Assessment

The *Teacher Toolkit* is the most important part of the FOSS Program. It is here that all the wisdom and experience contributed by hundreds of educators has been assembled. Everything we know about the content of the module, how to teach the subject, and the resources that will assist the effort are presented here.

FOSS Program Goals

FOSS Program Goals

Contents

Introduction 1
Goals of the FOSS Program 2
Bridging Research
into Practice 5
FOSS Next Generation K–8
Scope and Sequence 8

INTRODUCTION

The Full Option Science System™ has evolved from a philosophy of teaching and learning at the Lawrence Hall of Science that has guided the development of successful active-learning science curricula for more than 40 years. The FOSS Program bridges research and practice by providing tools and strategies to engage students and teachers in enduring experiences that lead to deeper understanding of the natural and designed worlds.

Science is a creative and analytic enterprise, made active by our human capacity to think. Scientific knowledge advances when scientists observe objects and events, think about how they relate to what is known, test their ideas in logical ways, and generate explanations that integrate the new information into understanding of the natural world. Engineers apply that understanding to solve real-world problems. Thus, the scientific enterprise is both what we know (content knowledge) and how we come to know it (practices). Science is a discovery activity, a process for producing new knowledge.

The best way for students to appreciate the scientific enterprise, learn important scientific and engineering concepts, and develop the ability to think well is to actively participate in scientific practices through their own investigations and analyses. FOSS was created to engage students and teachers with meaningful experiences in the natural and designed worlds.

Full Option Science System   Copyright © The Regents of the University of California

FOSS Program Goals

GOALS OF THE FOSS PROGRAM

FOSS has set out to achieve three important goals: scientific literacy, instructional efficiency, and systemic reform.

Scientific Literacy

FOSS provides all students with science experiences that are appropriate to students' cognitive development and prior experiences. It provides a foundation for more advanced understanding of core science ideas which are organized in thoughtfully designed learning progressions and prepares students for life in an increasingly complex scientific and technological world.

The National Research Council (NRC) in *A Framework for K–12 Science Education* and the American Association for the Advancement of Science (AAAS) in *Benchmarks for Scientific Literacy,* have described the characteristics of scientific literacy:

- Familiarity with the natural world, its diversity, and its interdependence.

- Understanding the disciplinary core ideas and the cross-cutting concepts of science, such as patterns; cause and effect; scale, proportion, and quantity; systems and system models; energy and matter—flows, cycles, and conservation; structure and function; and stability and change.

- Knowing that science and engineering, technology, and mathematics are interdependent human enterprises and, as such, have implied strengths and limitations.

- Ability to reason scientifically.

- Using scientific knowledge and scientific and engineering practices for personal and social purposes.

The FOSS Program design is based on learning progressions that provide students with opportunities to investigate core ideas in science in increasingly complex ways over time. FOSS starts with the intuitive ideas that primary students bring with them and provides experiences that allow students to develop more sophisticated understanding as they grow through the grades. Cognitive research tells us that learning involves individuals in actively constructing schemata to organize new information and to relate and incorporate the new understanding into established knowledge. What sets experts apart from novices is that experts in a discipline have extensive knowledge that is effectively organized into structured schemata to promote thinking. Novices have disconnected ideas about a topic that are difficult to retrieve and use. Through internal processes to establish schemata and through social processes of interacting with peers and adults, students construct understanding of the natural world and their relationship to it. The target goal for FOSS students is to know and use scientific explanations of the natural world and the designed world; to understand the nature and development of scientific knowledge and technological capabilities; and to participate productively in scientific and engineering practices.

Instructional Efficiency

FOSS provides all teachers with a complete, cohesive, flexible, easy-to-use science program that reflects current research on teaching and learning, including student discourse, argumentation, writing to learn, and reflective thinking, as well as teacher use of formative assessment to guide instruction. The FOSS Program uses effective instructional methodologies, including active learning, scientific practices, focus questions to guide inquiry, working in collaborative groups, multisensory strategies, integration of literacy, appropriate use of digital technologies, and making connections to students' lives, including the outdoors.

FOSS is designed to make active learning in science engaging for teachers as well as for students. It includes these supports for teachers:

- Complete equipment kits with durable, well-designed materials for all students.
- Detailed *Investigations Guide* with science background for the teacher and focus questions to guide instructional practice and student thinking.
- Multiple strategies for formative assessment at all grade levels.
- Benchmark assessments (grades 1–5) with online access for administering, coding, and analyzing assessments (grades 3–5).
- Strategies for use of science notebooks for novice and experienced users.
- *FOSS Science Resources,* a book of module-specific readings with strategies for science-centered language development.
- The FOSS website with interactive multimedia activities for use in school or at home, suggested interdisciplinary-extension activities, and extensive online support for teachers, including teacher prep videos.

FOSS Program Goals

FOSS Program Goals

Systemic Reform

FOSS provides schools and school systems with a program that addresses the community science-achievement standards. The FOSS Program prepares students by helping them acquire the knowledge and thinking capacity appropriate for world citizens.

The FOSS Program design makes it appropriate for reform efforts on all scales. It reflects the core ideas to be incorporated into the next-generation science standards. It meets with the approval of science and technology companies working in collaboration with school systems, and it has demonstrated its effectiveness with diverse student and teacher populations in major urban reform efforts. The use of science notebooks and formative-assessment strategies in FOSS redefines the role of science in a school—the way that teachers engage in science teaching with one another as professionals and with students as learners, and the way that students engage in science learning with the teacher and with one another. FOSS takes students and teachers beyond the classroom walls to establish larger communities of learners.

BRIDGING RESEARCH INTO PRACTICE

The FOSS Program is built on the assumptions that understanding core scientific knowledge and how science functions is essential for citizenship, that all teachers can teach science, and all that students can learn science. The guiding principles of the FOSS design, described below, are derived from research and confirmed through FOSS developers' extensive experience with teachers and students in typical American classrooms.

Understanding of science develops over time. FOSS has elaborated learning progressions for core ideas in science for kindergarten through grade 6. Developing the learning progressions involves identifying successively more sophisticated ways of thinking about core ideas over multiple years. "If mastery of a core idea in a science discipline is the ultimate educational destination, then well-designed learning progressions provide a map of the routes that can be taken to reach that destination" (National Research Council, *A Framework for K–12 Science Education*, 2012).

Focusing on a limited number of topics in science avoids shallow coverage and provides more time to explore core science ideas in depth. Research emphasizes that fewer topics experienced in greater depth produces much better learning than many topics briefly visited. FOSS affirms this research. FOSS modules provide long-term engagement (9–10 weeks) with important science ideas. Furthermore, modules build upon one another within and across each strand, progressively moving students toward the grand ideas of science. The core ideas of science are difficult and complex, never learned in one lesson or in one class year.

FOSS Next Generation—Elementary Module Sequences

	PHYSICAL SCIENCE		EARTH SCIENCE		LIFE SCIENCE	
	MATTER	ENERGY AND CHANGE	DYNAMIC ATMOSPHERE	ROCKS AND LANDFORMS	STRUCTURE/ FUNCTION	COMPLEX SYSTEMS
5	Mixtures and Solutions		Earth and Sun		Living Systems	
4		Energy		Soils, Rocks, and Landforms	Environments	
3	Motion and Matter		Water and Climate		Structures of Life	
2	Solids and Liquids			Pebbles, Sand, and Silt	Insects and Plants	
1		Sound and Light	Air and Weather		Plants and Animals	
K	Materials and Motion		Trees and Weather		Animals Two by Two	

FOSS Program Goals

FOSS Program Goals

Science is more than a body of knowledge. How well you think is often more important than how much you know. In addition to the science content framework, every FOSS module provides opportunities for students to engage in and understand science practices, and many modules explore issues related to engineering practices and the use of natural resources. FOSS promotes these science and engineering practices described in *A Framework for K–12 Science Education*.

- Asking questions (for science) and defining problems (for engineering)
- Developing and using models
- Planning and carrying out investigations
- Analyzing and interpreting data
- Using mathematics and computational thinking
- Constructing explanations (for science) and designing solutions (for engineering)
- Engaging in argument from evidence
- Obtaining, evaluating, and communicating information

Science is inherently interesting, and children are natural investigators. It is widely accepted that children learn science concepts best by doing science. Doing science means hands-on experiences with objects, organisms, and systems. Hands-on activities are motivating for students, and they stimulate inquiry and curiosity. For these reasons, FOSS is committed to providing the best possible materials and the most effective procedures for deeply engaging students with scientific concepts. FOSS students at all grade levels investigate, experiment, gather data, organize results, and draw conclusions based on their own actions. The information gathered in such activities enhances the development of scientific and engineering practices.

Education is an adventure in self-discovery. Science provides the opportunity to connect to students' interests and experiences. Prior experiences and individual learning styles are important considerations for developing understanding. Observing is often equated with seeing, but in the FOSS Program all senses are used to promote greater understanding. FOSS evolved from pioneering work done in the 1970s with students with disabilities. The legacy of that work is that FOSS investigations naturally use multisensory methods to accommodate students with physical and learning disabilities and also to maximize information gathering for all students. A number of tools, such as the FOSS syringe and balance, were originally designed to serve the needs of students with disabilities.

Formative assessment is a powerful tool to promote learning and can change the culture of the learning environment. Formative assessment in FOSS creates a community of reflective practice. Teachers and students make up the community and establish norms of mutual support, trust, respect, and collaboration. The goal of the community is that everyone will demonstrate progress and will learn and grow.

Science-centered language development promotes learning in all areas. Effective use of science notebooks can promote reflective thinking and contribute to life long learning. Research has shown that when language-arts experiences are embedded within the context of learning science, students improve in their ability to use their language skills. Students are eager to read to find out information, and to share their experiences both verbally and in writing.

Experiences out of the classroom develop awareness of community. By extending classroom learning into the outdoors, FOSS brings the science concepts and principles to life. In the process of validating classroom learning among the schoolyard trees and shrubs, down in the weeds on the asphalt, and in the sky overhead, students will develop a relationship with nature. It is our relationship with natural systems that allows us to care deeply for these systems.

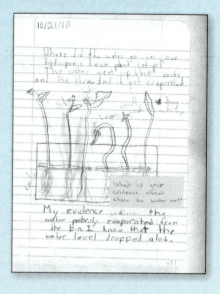

FOSS Program Goals

FOSS Program Goals

FOSS NEXT GENERATION K–8 SCOPE AND SEQUENCE

Grade	Physical Science	Earth Science	Life Science
6–8	Waves* Gravity and Kinetic Energy*	Planetary Science	Heredity and Adaptation* Human Systems Interactions*
	Chemical Interactions	Earth History	Populations and Ecosystems
	Electromagnetic Force* Variables and Design*	Weather and Water	Diversity of Life
5	Mixtures and Solutions	Earth and Sun	Living Systems
4	Energy	Soils, Rocks, and Landforms	Environments
3	Motion and Matter	Water and Climate	Structures of Life
2	Solids and Liquids	Pebbles, Sand, and Silt	Insects and Plants
1	Sound and Light	Air and Weather	Plants and Animals
K	Materials and Motion	Trees and Weather	Animals Two by Two

* Half-length course

FOSS is a research-based science curriculum for grades K–8 developed at the Lawrence Hall of Science, University of California, Berkeley. FOSS is also an ongoing research project dedicated to improving the learning and teaching of science. The FOSS project began over 25 years ago during a time of growing concern that our nation was not providing young students with an adequate science education. The FOSS Program materials are designed to meet the challenge of providing meaningful science education for all students in diverse American classrooms and to prepare them for life in the 21st century. Development of the FOSS Program was, and continues to be, guided by advances in the understanding of how people think and learn.

With the initial support of the National Science Foundation and continued support from the University of California, Berkeley, and School Specialty, Inc., the FOSS Program has evolved into a curriculum for all students and their teachers, grades K–8. The current editions of FOSS are the result of a rich collaboration among the FOSS/Lawrence Hall of Science development staff; the FOSS product development team at School Specialty; assessment specialists, educational researchers, and scientists; and dedicated professionals in the classroom and their students, administrators, and families.

We acknowledge the thousands of FOSS educators who have embraced the notion that science is an active process, and we thank them for their significant contributions to the development and implementation of the FOSS Program.

Science Notebooks in Grades 3-5

Science Notebooks in Grades 3–5

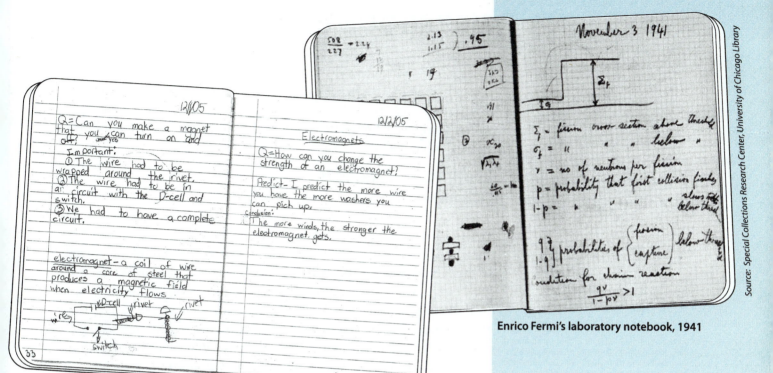

Student notebook from the Energy Module

Enrico Fermi's laboratory notebook, 1941

INTRODUCTION

A scientist's notebook is a detailed record of his or her engagement with natural phenomena. It is a personal representation of experiences, observations, and thinking—an integral part of the process of doing scientific work. A scientist's notebook is a continuously updated history of the development of scientific knowledge and reasoning. FOSS students are young scientists; they incorporate notebooks into their science learning.

This chapter is designed to be a resource for teachers who are incorporating notebooks into their classroom practice. For teachers just beginning to use notebooks, the Getting Started section in this chapter suggests how to set up the notebooks, and the *Investigations Guide* cues you when to engage students with the notebooks during the investigation. For more information on specific types of notebook entries, the subsections in the Notebook Components section include strategies to differentiate instruction for various ability levels.

Contents

Introduction	1
Notebook Benefits	2
Getting Started	7
Notebook Components	12
Planning the Investigation	12
Data Acquisition and Organization	15
Making Sense of Data	18
Next-Step Strategies	22
Writing Outdoors	26
Closing Thoughts	28

FOSS Full Option Science System

Science Notebooks in Grades 3–5

NOTEBOOK BENEFITS

Engaging in active science is one part experience and two parts making sense of the experience. Science notebooks help students with the sense-making part. Science notebooks assist with documentation and cognitive engagement. For teachers, notebooks are tools for gaining insight into students' thinking. Notebooks inform and refine instructional practice.

Benefits to Students

Benefits to Students
- *Documentation*
- *Reference document*
- *Cognitive engagement*

Documentation. Science provides an authentic experience for students to develop their documentation skills. Students are encouraged to keep science notebooks to organize and learn to communicate their thinking. They document their experiences, data, and thinking during each investigation. Students create tables, graphs, charts, drawings, and labeled illustrations as standard means for representing and displaying data. At first, students will look at their science notebooks as little more than a random collection of words and pictures. Each notebook page represents an isolated activity. As students become more accomplished at keeping notebooks, their documentation will become better organized and efficient. In time and with some guidance, students will adopt a deeper understanding of their collections as integrated records of their learning.

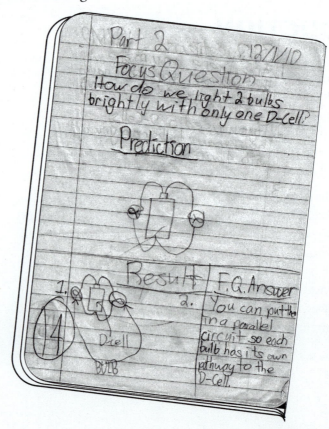

Science notebook entry from the Energy Module

Reference document. When data are displayed in functional ways, students can think about them more effectively. A well-kept notebook is a useful reference document. When students have forgotten a fact about a rock that they learned earlier in their studies, they can look it up. Learning to trust a personal record of previous discoveries and knowledge structures is important.

A complete and accurate record allows students to reconstruct the sequence of learning events to "relive" the experience. Discussions about science among students; students and teachers; or students, teachers, and families have more meaning when they are supported by authentic documentation in students' notebooks.

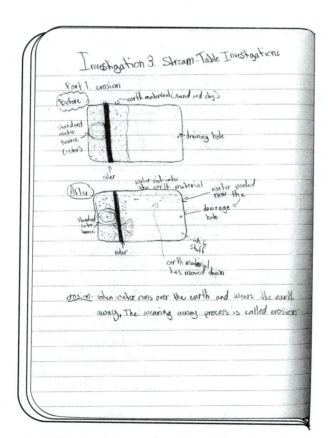

From the Soils, Rocks, and Landforms Module

Science Notebooks in Grades 3–5

Cognitive engagement. Once data are recorded and organized in an efficient manner in science notebooks, students can think about them to draw conclusions about the way the world works. Their data, based on their experiences and observations, are the raw materials that students use to forge concepts and relationships.

Writing stimulates active reasoning. There is a direct relationship between the formation of concepts and the rigors of expressing those thoughts in words. Writing requires students to impose discipline on their thoughts. By writing explanations, students clarify what they know . . . and expose what they don't know.

When students are asked to generate derivative products (such as detailed explanations or posters) as evidence of learning, the process will be much more efficient when they have a coherent, detailed notebook for reference. As students begin to use their notebooks as a personal reference text, they will value their own learning and come to rely on their own work as a source of information about science.

When students use their notebooks as an integral part of their science studies, they are more likely to think critically about their understandings. This reflective thinking can be encouraged by notebook entries that present opportunities for self-assessment. Self-assessment motivates students to rethink and restate their scientific understandings. Revising their notebook entries by using a next-step strategy helps students clarify their understanding of the science concepts under investigation.

▶ **NOTE**
For more on derivative products, see the Science-Centered Language Development chapter.

Benefits to Teachers

In FOSS, the unit of instruction is the module—a sequence of conceptually related learning experiences that leads to a set of learning outcomes. A science notebook helps you think about and communicate the conceptual structure of the module you are teaching.

Assessment. From the assessment point of view, a science notebook is a collection of student-generated artifacts that exhibit learning. You can assess student skills, such as using drawings to record data, while students are working with materials. At other times, you collect the notebooks and review them in greater detail. The displays of data and analytical work, such as responses to focus questions, provide a measure of the quality and quantity of student learning. The notebook itself should not be graded. However, the notebook can be considered as one component of a student's overall performance in science.

Medium for feedback. The science notebook is an excellent medium for providing feedback to individual students regarding their work. Most students will be able to read a teacher comment written on a self-stick note, think about the issue, and respond. Some students may need oral feedback individually or in a small-group situation. This feedback might include modeling to help students make more accurate drawings, revisiting some essential scientific vocabulary, or introducing strategies to better explain their thinking.

Focus for professional discussions. The science notebook acts as a focal point for discussion about students' learning at several levels. It can be reviewed and discussed during parent conferences. Science notebooks can be the focus of three-way discussions among students, teachers, and principals to ensure that all members of the school science community agree about what kinds of student work are valued and the level of performance to expect. Science notebooks shared among teachers in a study group or other professional-development environment can serve as a reflective tool that informs teachers of students' ability to demonstrate recording techniques, individual styles, various levels of good-quality work, and so on. Just as students can learn notebook strategies from one another, teachers can learn notebook skills from one another.

Benefits to Teachers
- *Assessment*
- *Medium for feedback*
- *Focus for professional discussions*
- *Refinement of practice*

Science Notebooks in Grades 3–5

Refinement of practice. As teachers, we are constantly looking for ways to improve instructional practices to increase students' understanding. Your use of the notebook should change over time. In the beginning, the focus will be on the notebook itself—what it looks like, what goes in it. As you become more comfortable with the notebook, the attention shifts to what students are learning. When this happens, you begin to consider how much scaffolding to provide to different students, how to use evidence of learning to differentiate instruction, and how to modify instruction to refine students' understanding.

GETTING STARTED

A major goal for using notebooks is to establish habits that will enable students to collect data and make sense of them. Use of the notebook must be flexible enough to allow students room to grow and supportive enough for students to be successful from the start. The format should be simple and the information meaningful to students. The notebook includes student drawings; writing in the form of individual words, short phrases, and sentences; and a variety of visual and tactile artifacts. When students thumb through their notebooks, they should be reminded of the objects and organisms they observed and their interactions with them.

Notebook Format

The *Teacher Resources* component of the FOSS *Teacher Toolkit* includes duplication masters for the same notebook sheets that are bound into the consumable notebooks. You can use these sheets to prepare an analogous notebook or to have students design a customized version in a composition book.

In an autonomous approach, students create their entire science notebooks from blank pages in bound composition books. Experienced students determine when to use their notebooks, how to organize space, what methods of documentation to use, and how to flag important information. This level of notebook use will not be realized quickly; it will likely require systematic development by an entire teaching staff over several years.

You might choose to have a separate notebook for each module or one notebook for the entire year. (See the sidebar for the advantages of each.) Students will need about 30 pages (60 sides) for a typical module.

Advantages of One Notebook per Module
- *Easy to replace if lost*
- *Lower cost*
- *Fewer pages*

Advantages of One Notebook per Year
- *Easy to refer to prior activities*
- *Easy to see growth over time*

Science Notebooks in Grades 3–5

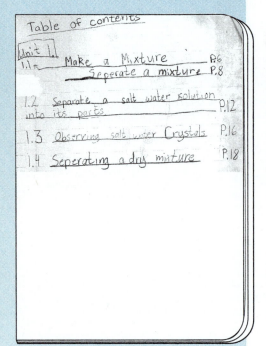

A sample table of contents

Organization of Notebooks

Four organizational components of the notebook should be planned right from the outset: a table of contents, page numbering, documentation, and an index. The consumable FOSS science notebook has these features already in place. No setup is required beyond pointing out where these features are in the notebook.

Table of contents. Students should reserve the first two pages of their notebooks for a table of contents. They will add to it systematically as they proceed through the module. The table of contents can be based on the names of the investigations in the module, the specific activities undertaken, the concepts learned, or some other schema that makes sense to everyone. You could have students make their own titles for the table of contents. This is best done at the end of the investigation part so that students can come up with meaningful titles.

Page numbering. Each page should have a number. These can be applied to the pages and referenced in the table of contents as the notebook progresses, or small blocks of pages can be prenumbered (pages 1–15 initially, pages 16–30 later, and so on) at appropriate times in the module.

Date, title, and other conditions. Each time students make a new entry, they should record certain information. At the very minimum, they should record the date and a title. More complete documentation might include the time; day of the week; team members; and, if appropriate, weather conditions.

Some classes start each new entry at the top of the next available page. Others simply leave a modest space and apply the documentation information right before the new entry.

When introducing a new condition to students, such as weather conditions, it is important to discuss why students are recording the information so that they understand its relevancy.

Index. Scientific academic language is important. FOSS strives to have students use precise, accurate vocabulary at all times in their writing and conversations. To assist with acquisition of the scientific vocabulary, students should set up an index at the end of their notebooks. It is not usually possible to enter the words in alphabetical order, since they will be acquired as the module advances. Instead, assign a block of letters to each of several index pages (A–E, F–K, etc.) where students can enter key words. As an alternative, students could use a single page blocked out in 24 squares and assign one or more letters to each square. (See the sample illustration.) Students write the new vocabulary word or phrase in the appropriate square and tag it with the number of the page on which the word is defined in the notebook.

Notebook Entries

As students engage in scientific exploration, they will make entries in their notebooks. They may use a prepared notebook sheet or create a more free-form entry. Students frequently respond to a focus question with a drawing or a simple narrative entry. For example, in the **Soils, Rocks, and Landforms Module**, students may explain their thinking about factors that affect erosion and include a drawing to show how the earth materials move when the land is steep. These notebook entries allow students to relive and describe their science experiences as they turn the pages in their notebooks.

Typically, the rules of grammar and spelling are relaxed when making notebook entries so as not to inhibit the flow of creative expression. Encourage students to use many means of recording and communicating besides writing, including charts, graphs, drawings, graphics, color codes, numbers, and images attached to the notebook pages. By exploring many options for making notebook entries, each student will find his or her most efficient, expressive way to capture and organize information for later retrieval.

▶ **NOTE**
Templates for indexes can be found online at www.FOSSweb.com.

A sample index

▶ **NOTE**
More information about vocabulary appears in the Science-Centered Language Development chapter.

Science Notebooks in Grades 3–5

Supporting Students

Elementary classrooms contain students with a range of abilities, which need to be considered when planning strategies for implementing science notebooks. Students need to have successful early experiences with notebooks. A blank notebook may be intimidating for some students, and they will look to you for guidance. FOSS teachers have had success using different supportive structures to help transform the blank notebook into a valuable reference tool. These supports and scaffolds can be used with the whole class, a small group, or an individual and should be adjusted to meet students' needs.

Class notebook. You can create a class notebook to document the investigation as a way to model the various notebook components. The class notebook should be accessible at all times for students to reference. You can use a chart-paper tablet, a paper notebook displayed under a document camera, or a computer. You could use the class notebook to introduce strategies such as a T-table, or write a summary statement in it after all students have answered the focus question. While individual notebooks will look similar to the class notebook, it is not the intent that students' notebooks be identical to the class notebook.

Scaffolds. Supports and scaffolds differ in one way. Supports are always available for students to access, such as allowing students access to a class notebook. Scaffolds are available just when the student needs them and will vary from student to student and from investigation to investigation. Scaffolds are meant to provide structure to a notebook entry and allow students to insert their observations or thinking. As the year progresses, the scaffolds change to allow for more student independence. Scaffolds include

- **Sentence starters** or **drawing starters** provide a beginning point for a notebook entry. Here's an example: "I noticed the crayfish had _____."

- **Frames** provide more support but leave specific gaps for students to complete. Here's an example: "Magnets attract when _____. Magnets repel when _____."

- The suggested **notebook sheets** can guide students to collect data with a table, graph, or list of questions. The notebook sheets also guide thinking.

> **TEACHING NOTE**
>
> *Use of the class notebook should be thoughtfully timed. Doing a class-notebook entry at the end of an activity is helpful to teach the components of a notebook, yet allows you to see what students do on their own. If you want to model a specific notebook strategy, use the class notebook during the activity.*

Think-alouds. Think-alouds help explain the decision-making process practiced by a savvy notebook user. They verbalize the thoughts used to create a particular notebook entry. For example, if students have recorded observations about how one variable affects erosion and deposition and are going to observe a second variable, you might say,

I am going to observe how another variable affects erosion and deposition. I am going to look back to see how I recorded the information before. I see that I made a detailed drawing. I used a T-table to describe what happened to the earth materials over time. So now I'm going to make similar observations and use a T-table and then make a drawing. One way I can get ideas about how to organize my observations is to look back at observations I wrote before.

Providing time to record. When students are engaged in active science, their efforts are focused on the materials, not the notebook. Students need this time to explore and initially might not open their notebooks and record observations until prompted. Students may need separate time to record observations that fully document their discoveries. Some teachers have found it easier to leave the materials on the table and have students bring their notebooks to a common writing area. Then the teacher revisits the focus question or task and provides a few minutes for students to record in their notebooks.

Ownership

A student's science notebook can be personal or public. If the notebook is personal, the student decides how accessible his or her work is to other students. If ownership falls at the opposite extreme, everything is public, and anyone can look at the contents of anyone else's notebook at any time. In practice, most classroom cultures establish a middle ground in which a student's notebook is substantially personal, but the teacher claims free access to the students' work and can request that students share notebooks with one another and with the whole class from time to time.

> **TEACHING NOTE**
>
> *Consider having students who need extra assistance with writing or formulating their ideas dictate specific information to an adult. The adult writes the information in the notebook for the student. Or the adult could write the sentence, using a highlighter, and the student could trace the words, using a pencil.*

Science Notebooks in Grades 3–5

Notebook Components
- *Planning the investigation*
- *Data acquisition and organization*
- *Making sense of data*
- *Next-step strategies*

NOTEBOOK COMPONENTS

A few components give the science notebook conceptual shape, coherence, and direction. These components don't prescribe a step-by-step procedure for how to prepare the notebook, but they do provide some overall guidance.

The general arc of an investigation starts with a question or challenge, and then proceeds with an activity, data acquisition, sense making, and next steps. The science notebook should record important observations and thoughts along the way. It may be useful to keep these four components in mind as you systematically guide students through their notebook entries.

Planning the Investigation

Typically at the start of a new activity, the first notebook entry is a focus question, which students transcribe into their notebooks. The focus question establishes the direction and conceptual challenge for the investigation.

Focus question. Each part of each investigation starts with a focus question or challenge. Write or project it on the board for students to transcribe into their notebooks, or give them photocopied strips of the focus question to tape or glue into their notebooks. Some teachers look ahead, write all the focus questions on one sheet of paper, copy the sheet, and cut the questions apart with a paper cutter. A list of the focus questions for each module is also available as a PDF on FOSSweb.

> 2. **Focus question: What's in our schoolyard soils?**
>   Write or project the focus question on the board, and say it aloud.
>   ➤ *What's in our schoolyard soils?*

Students may develop their own questions to investigate as well.

The focus question determines the kinds of data to be collected and the procedures that will yield those data. The procedures may be a narrative plan consisting of a few sentences or a more detailed step-by-step procedure, depending on the requirements of the investigation.

Narrative plans. After posing a focus question, we often ask students for their ideas about how they will engage the question. For instance, the focus question that sets up a free exploration of crayfish might be

➤ *What do crayfish do when they are removed from water for a short time?*

Students think about this question and formulate a plan, perhaps writing a short narrative description of their general approach.

We are going to take the crayfish out and put it on the table for 3 minutes. We are going to see if the swimmerets move the same way they do in the water.

The narrative helps students consider the options available to them. It also reminds them of the limits and considerations when working with living organisms.

As with all notebook entries, some students need little more than a nudge in the right direction. Other students may need more support. One option is to prompt them with sentence starters. Good sentence starters provide a start but still leave the intellectual responsibility with students. Here are some examples.

- *The first thing we are going to do is _____.*
- *The next thing we will do is _____.*
- *We have to be careful about _____.*
- *Finally, we are going to _____.*

Lists. Science notebooks often include lists of things to think about, materials to get, or words to remember. A materials list is a good organizer, helping students anticipate actions they will take. A list of variables to be controlled clarifies the purpose of an experiment. Simple lists of dates for making observations, or of the people responsible for completing a task keep information readily available.

> **TEACHING NOTE**
>
> *Supports such as sentence starters and frames should be monitored and adjusted as your expectation of students' responsibility for notebooks changes over the year.*

Science Notebooks in Grades 3–5

TEACHING NOTE

Be selective when engaging students in making step-by-step procedures as the process will require additional time. To save time when using a class-generated procedure, provide each student with a typed copy.

TEACHING NOTE

It is important for students to understand that in order to do accurate work, scientists need to do multiple trials and present their work so that it can be replicated.

Step-by-step procedures. Some work with materials requires structured planning. Students start an investigation in the **Soils, Rocks, and Landforms Module** with the focus question

➤ *What's in our schoolyard soils?*

Students need to recall what they know about soil and develop a strategy for adding water to help isolate the different components. An appropriate convention for recording a sequential procedure is a numbered, step-by-step plan. Here is an example.

1. Collect 1 spoon of soil.
2. Record where it came from.
3. Put it in a vial.
4. Add 20 mL of water.
5. Cap the vial and shake.
6. Let it settle overnight.
7. Observe layers.

One way to introduce students to this type of entry is to provide a notebook sheet as a model. Both you and students can refer to the notebook sheet as they work through the hands-on investigation. During the next investigation, students can look back at the model notebook sheet when they write their own step-by-step procedures. To check the procedures for errors or omissions, students can trade notebooks and attempt to follow other students' instructions to complete the task.

Predictions. Depending on the content and the focus question, students may be able to make a prediction. When they make predictions, they are attempting to relate prior experiences to the question posed. Providing students with a frame can help them explain the rationale behind their predictions. A frame to help with stating a prediction is "I think that _____ because _____."

Data Acquisition and Organization

Data are the bits of information (observations) from which scientists construct ideas about the structure and behaviors of the natural world. Because observation is the starting point for answering the focus question, data records should be

- clearly related to the focus question;
- accurate and precise, including units with measurements;
- organized for efficient reference.

Data handling can have two phases: data acquisition and data display. Data acquisition is making observations and recording data. The data record can be composed of words, phrases, numbers, and/or drawings. Data display is reorganizing the data in a logical way to facilitate thinking. The display can take the form of narratives, drawings, artifacts, charts, tables, images, graphs, Venn diagrams, calendars, or other graphic organizers. Early in a student's experience with notebooks, the record may be disorganized and incomplete, and the display may need guidance. With practice, however, students will become skilled at determining what form of recording to use in various situations and how best to display the data for analysis.

Narratives. The most intuitive approach to recording data for most students is narrative—using words, sentence fragments, and numbers in a more or less sequential manner. As students make a new observation, they record it right after the previous entry, followed by the next observation, and so on. Some observations, such as a record of the movements and interactions of a crayfish over time or the changes observed in an environment, are appropriately recorded in narrative form.

Sentence frames can help students record their narratives initially. Another option is to have students read their narratives to a partner from a different group. This allows them to refine their language and gain insight into another student's observation.

Science Notebooks in Grades 3–5

Science Notebooks in Grades 3–5

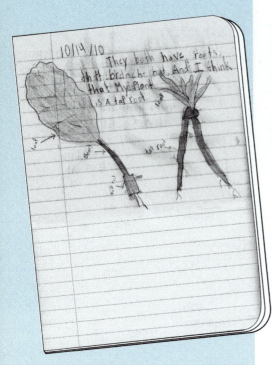

A drawing from the Structures of Life Module

Drawings. When students observe the shape and distribution of crystals after evaporation, or observe and identify the parts of a system, a labeled illustration is the most efficient way to record data. A picture is worth a thousand words, and a labeled picture is even more useful.

Students' initial drawings may need refinement to accurately capture their observations. It can be helpful to suggest an acronym for making useful drawings. Accurate, big, colorful, and detailed (ABCD) drawings can explain systems, identify structures of an organism, or capture observations of landforms. As the need arises, introduce specific drawing techniques, such as the use of scale, magnified view, and appropriate use of color.

Tables. With experience, students will recognize when a table is appropriate for recording data. When students make similar observations about a series of objects, such as a set of minerals or powdered substances, a table with columns is an efficient recording method. The two-dimensional table makes it easy to compare the properties of all the objects under investigation. Similarly, when students conduct an experiment, they can record data directly into a T-table. With little effort, they can transform the table, presented in ordered pairs, into a graph.

Notebook sheets are a good way to scaffold the use of tables. Students can quickly enter the information on the sheets. As they take on more independence in their notebooks, discussions about the column headings or the purpose of the tables shift the focus from filling in the table, to the purpose the table serves, and, eventually, to how to make a table to record this information.

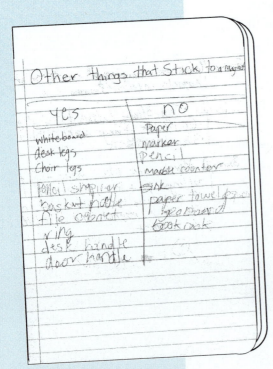

A table from the Energy Module

Graphs, charts, and graphics. Reorganizing data into a logical, easy-to-use graphic is the first phase of data analysis. Graphs allow easy comparison (bar graphs), quick statistical analysis of numerical data (line plots), and visual confirmation of a relationship between variables (two-coordinate graphs). Additional graphic tools, such as Venn diagrams, pie charts, concept maps, food chains, and life cycles, help students perceive patterns in their data.

As students progress in making graphs and graphics, so should their understanding of when and why to use them. Taking a little time to have students discuss which graph to use before distributing a notebook sheet will go a long way to reinforce the idea that different graphs serve different purposes.

Artifacts. Occasionally, the results of an investigation produce three-dimensional products that students can tape or glue directly into their science notebooks. Sand, minerals, seeds, and so on can become a permanent part of the record of learning.

Images. Digital photos of plants, rocks, and the results of investigations can be great additions to the science notebook. Digital photos should be used to enhance students' observations and drawings, not to replace them.

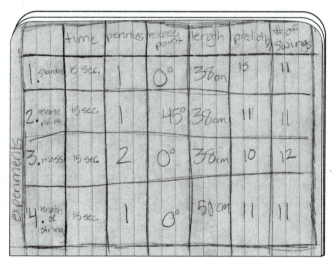

A table from the Motion, Force, and Models Module

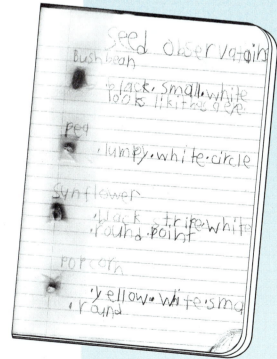

Artifacts can be placed inside the notebook.

Science Notebooks in Grades 3–5

Science Notebooks in Grades 3–5

Making Sense of Data

Most of the sense making takes place during whole-group discussions when students share and discuss the observations they made while investigating. Questions are asked to help students interpret and analyze data in order to build conceptual understanding. Encourage students to use new vocabulary when they are sharing and making sense of their data. These words should be accessible to all students on the word wall. When students have limited written language skills, this oral discussion is important. This sharing is essential and is described in detail in the *Investigations Guide*. After sharing observations, the class revisits the focus question. Students flip back in their notebooks and use their data to discuss their answers to the question. You scaffold the discussion, and use appropriate language-development strategies.

The student's most important job is to think about the data and the focus question together to craft an answer to the question. Many investigations give students time to discuss and think with fellow students prior to writing an answer.

---

14. Discuss the relative amounts of materials in soils

Have students look at other groups' vials and the vials of soils you saved from Part 1 in order to compare the contents.

➤ *Do all vials contain the same amount of soil?* [Approximately.]

➤ *Do all vials have the same number of layers?* [Possibly, but not necessarily.]

➤ *Are the layers the same thickness for each type of earth material?* [Probably not.]

➤ *Can you identify each layer?* [Each soil should include some sand, a significant layer of silt and clay, and some humus.]

➤ *How are our schoolyard soils alike and different from the mountain, desert, river delta, and forest soils?*

After the sense-making discussion, students might review relevant vocabulary. They answer the focus question in their notebooks, expressing their current understanding. Some students will naturally use scientific vocabulary in their explanations, while others may need additional experiences before they include scientific vocabulary in their writing. In many instances, you can use supports to guide the development of a coherent and complete response to the question.

Frames and prompts. When providing frames or prompts, you are helping students organize their thinking. The frame provides a communication structure that allows students to focus their attention on thinking about the science involved. Here are some general frames.

- *I used to think _____, but now I think _____.*
- *The most important thing to remember about evaporation is _____.*
- *One thing I learned about _____.*

> **TEACHING NOTE**
> Frames and prompts are designed to support students to either communicate or think more about the content. The important point is that students are doing the thinking and not merely responding to the prompt.

When using frames, you need to provide just the right amount of scaffolding to allow students to communicate their understanding. Too much scaffolding limits the opportunity for students' communication skills to progress. Too little scaffolding may not allow the student to accurately represent his or her understanding.

For example, in the **Energy Module**, students are asked,

➤ *What is the relationship between the distance separating two magnets and the force of attraction between them?*

You can provide different frames to students, depending on ability. Each frame has merit when used appropriately. The four different frames below give students progressively less support.

- *As the distance between the two magnets increases, the force of attraction _____. I know this because when I tested a distance of 6 spacers, the force broke with _____ washers. When I tested a distance of 1 spacer, the force broke with _____ washers.*
- *As the distance between the two magnets _____, the force of attraction _____. I know this because when I tested a distance of _____ spacers, the force broke with _____ washers. When I tested a distance of _____ spacer, the force broke with _____ washers.*
- *As the _____, the _____. I know this because _____.*
- *Make a claim about the force of attraction between magnets, and provide evidence to support your claim.*

Science Notebooks in Grades 3–5

Science Notebooks in Grades 3–5

Claims and evidence. A claim is an assertion about how the natural world works. A student might claim, for instance, that metals stick to magnets. For the claim to be valid and accurate, it must be supported by evidence—statements that are directly correlated with data. The evidence should refer to specific observations, relationships that are displayed in graphs, tables of data that show trends or patterns, dates, measurements, and so on. Many teachers will provide a frame to help students include both their claims and evidence in their responses. A claims–and–evidence construction is a sophisticated, rich display of student learning and thinking.

Conclusions and predictions. At the end of an investigation (major conceptual sequence), it may be appropriate for students to generate a summarizing narrative to succinctly communicate what they have learned. When appropriate, students can make predictions based on their understanding of a principle or relationship. For instance, after completing the investigation on evaporation, a student might predict the order in which the water will completely evaporate from various containers, based on surface area exposed to air. Or a student might predict how long to make a pendulum that swings 15 times in 15 seconds. Predictions will frequently indicate the degree to which a student can apply the new knowledge to real-world situations. A prediction can be the springboard for further inquiry by the class or by individual students.

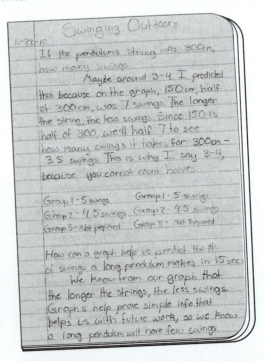

A prediction can show understanding of a relationship.

I wonder. Does the investigation connect to a student's personal interests? Or does the outcome suggest a question or pique a student's curiosity? Providing time for students to write "I wonder" statements or questions supports the idea that the pursuit of scientific knowledge does not end with the day's investigation. The notebook is an excellent place to capture students' musings and for students to record thoughts that might otherwise be lost.

Wrap-up/warm-up. At the end of each investigation part or at the beginning of the next part, students will engage in a wrap-up or warm-up. This is another opportunity for students to revisit the content of the investigation. As partners discuss their responses to the focus question, they may choose to add information to their responses. This should be encouraged.

21. Share notebook entries

Conclude Part 1 or start Part 2 by having students share notebook entries. Ask students to open their science notebooks to the most recent entry. Read the focus question together.

➤ *What is soil?*

Ask students to pair up with a partner to

- share their answers to the focus question;
- discuss the different kinds of materials that were found in the four different soils;
- share their ideas about which soil came from which location—desert, river delta, mountain, forest.

▶ **NOTE**
More information about partner and small-group discussion protocols that can be used during the Wrap-Up/Warm-Up session appears in the Science-Centered Language Development chapter.

Science Notebooks in Grades 3–5

Next-Step Strategies

In each investigation, the *Investigations Guide* indicates an assessment opportunity and what to look for when examining students' work. The purpose of looking at students' work at intermediate junctures is for embedded (formative) assessment, *not* for grading. Look for patterns in students' understanding by collecting and sorting the notebooks. If the patterns indicate that students need additional help with communication or with content, you might want to select a corrective next-step strategy before going on to the next part. This process of looking critically at students' work is described in more detail in the Assessment chapter.

A next-step strategy is a brief instructional activity that takes place before the start of the next investigation part. Select a strategy based on students' needs to communicate their thinking more efficiently or accurately, use scientific vocabulary effectively, or think about the concept in a different way.

During each next step, students are engaged in reflection and self-assessment. Scientists constantly refine and clarify their ideas about how the natural world works. They read scientific articles, consult with other scientists, and attend conferences. They incorporate new information into their thinking about the subject they are investigating. This reflective process can result in deeper understanding or even a complete revision in thinking.

Likewise, when students receive additional instruction or information after writing a conclusion, response sheet, or answer to a focus question, give them time to reflect on how their ideas might have changed. In the self-assessment that follows, they review their original written work, judge its accuracy and completeness, and write a revised explanation.

What follows here is a collection of next-step strategies that teachers have used successfully with groups of students to address areas of need. The strategies are flexible enough to use in different groupings and can be modified to meet your students' needs.

By engaging in any of these next-step strategies, students have to think actively about every aspect of their understanding of the concept and organize their thoughts into coherent, logical narratives. The learning that takes place during this reflection process is powerful. The relationships between the several elements of the concept become unified and clarified.

> **TEACHING NOTE**
>
> Students' learning can be assessed only at the level in which it was done. If students worked in groups to answer the focus question, it is difficult to assess individual understanding. Similarly, providing a frame to guide a student response provides evidence on what students can do at a supported level, not at an independent level.

Next-Step Strategies
- *Teacher feedback*
- *Line of learning*
- *Review and critique*
- *Key points*
- *Revision with color*
- *Class debate*
- *Mini-lessons*

Teacher feedback. Students' writing often exposes weaknesses in students' understanding—or so it appears. It is important to check whether the flaw results from poor understanding of the science or from imprecise communication. You can use the notebook to provide feedback, asking for clarification or additional information. Attach a self-stick note, which can be removed after the student has taken appropriate action. The most effective forms of feedback relate to the content of the work. Nonspecific feedback (such as stars, smiley faces, and "good job!") and ambiguous critiques (such as "try again," "put more thought into this," and "not enough") are less effective. Here are a couple of examples.

- *You claim that water condensed on the glass of ice water. Where did the water come from?*
- *Tell me why you think Lightbulb A will light. Hint: Check the contact points on the two illustrations.*

Feedback that guides students to think about the content of their work and gives suggestions for how to improve are productive instructional strategies. Here are some examples of useful generic feedback.

- *Use the science vocabulary in your answer.*
- *Include an example to support your ideas.*
- *Include more detail about _____.*
- *Check your data to make sure this is right.*
- *Include units with your measurements.*

When students return to their notebooks and respond to the feedback, you will have additional information to help you discriminate between knowledge and communication difficulties.

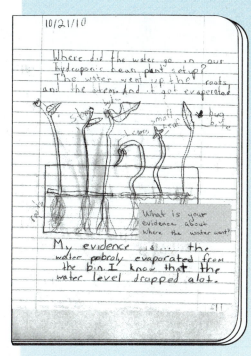

Students can respond to feedback using a different-color pencil.

Science Notebooks in Grades 3–5

Science Notebooks in Grades 3–5

Line of learning. One technique many teachers find useful in the reflective process is the line of learning. After students enter their initial explanation, followed by discussion, assessment, reading, and/or teacher feedback, they draw and date a line under their original work. Then they make a new entry under the line of learning, adding to or revising their original thinking. If the concept is elusive or complex, a second line of learning, followed by more processing and revising, may be appropriate.

The line of learning is a reminder to students that learning is an ongoing process, and that the products of the process are imperfect. The line of learning marks places in that process where a student made a stride toward full understanding. The psychological security provided by the line of learning reminds students that they can always draw another line of learning and revise their thinking once again. The ability to look back in the science notebook and see concrete evidence of learning gives students confidence and helps them become critical observers of their own learning.

Review and critique anonymous student work. Presenting work from other students can be a valuable learning tool for refining and improving responses for content and literacy. Depending upon the culture of the class, you might present actual or simulated student work from a focus question or response sheet, selected to represent a common misconception, error, or exemplary work. Present it to the whole class, and then have students work in groups to discuss the merits of the response and make recommendations for improvement. In this process, students discuss what information is needed in a quality response. After critiquing other students' responses, students look at their own responses, draw a line of learning, and refine their own thinking.

Key points. As students advance in years, so does the level of questions they are asked to answer. Some questions require students to address several points for a complete response. If students discuss only one point well, it may be that they are unaware of the other points or need assistance in how to address multiple points in an answer. Using key points, you pose the question to the group and, through discussion, elicit the key ideas or points that would be included in a complete answer to the question. On the board, record only words or brief phrases. Once the class has agreed on the key points, students review their responses and add information to answer the question completely. The list of words and phrases provides the support for revision, but not the language.

Revision with color. Suppose a student response of several sentences has some accurate information, some information that needs refinement, and a critical gap. Instead of crossing out the original response completely and starting anew, students can use color to revise their responses. After discussing the question with a partner or in a group, or even during key points, students grab three different-color pens or pencils and refine their responses by using the three Cs (confirm, correct, and complete). To confirm information as accurate, they underline it in green. If they need to correct a misconception, they do so in red. If they need to complete their responses by adding more information, they do it in blue. In doing so, students learn that their initial responses are works in progress, and, like scientists, students may revisit and modify their responses based on new information.

Class debate. If students have differing points of view on a response, you could have students engage in a healthy class debate. After establishing rules to foster a respectful and helpful class environment, volunteers read their answers, and classmates can agree or disagree. Students can present supporting evidence or counterarguments. An important part of the debate is that students may change their minds once they have additional evidence. Students return to their notebooks and can modify their responses by using any other next-step strategy.

Mini-lessons. Sometimes the data from sorting notebook entries reveal that students need some additional information or specific guidance on a skill. A mini-lesson is a brief interaction with a group of students, targeting the area of need. You might have a group of students observe a stream table more closely in order to observe landforms or give a writing prompt to a small group of students and ask them to explain their thinking more clearly.

Science Notebooks in Grades 3–5

WRITING OUTDOORS

Every time you go outdoors with students, you will have a slightly different experience. Naturally, the activity or task will be different, but other variables may change as well. The temperature, cloud cover, precipitation, moisture on the ground; other activities unexpectedly happening outside; students' comfort levels related to learning outdoors; and time of the school year are all aspects that could affect the activity and will certainly determine how you incorporate the use of notebooks. Most students are completely capable of staying on task with the outdoor experience when they are doing science, but writing outdoors can be a bit trickier. The following techniques are tried-and-true ways to help students learn how to write outdoors and to give them all the supplies they need to support their writing.

Create "Desks"

Students need a firm writing surface. Students who write in composition notebooks with firm covers can simply fold them open to the pages they are writing on, rest them in the crook of their nonwriting arms, and hold them steady with their nonwriting hands—they can stand, sit, kneel, or lean against a wall to write. At the beginning of the school year, take a minute to model how to do this.

Many students feel most comfortable sitting down to write. Curbs, steps, wooden stumps or logs, rocks, and grass are places to sit while writing. Select a writing location that suits your students' comfort levels. Some students will not be comfortable sitting on the grass or ground. They will need to sit on something such as a curb, stair, boulder, or stump at the start of the year, but will eventually feel more comfortable with all aspects of the outdoor setting as the year moves along.

If students are using individual notebook sheets or notebooks with flimsy covers, you will likely want to buy or make clipboards. If you do not have clipboards, use a box cutter to cut cardboard to the proper size. Clamp a binder clip at the top to make a lightweight yet sturdy clipboard. If it gets ruined, no tears will be shed. If you're in the market for new clipboards, get the kind that are stackable and do not have a bulky clip. Ideally, all the clipboards will fit in one bag for portability and easy distribution.

If using a notebook sheet, simply put the sheet on the clipboard before going outdoors, and have students glue the sheet into their notebooks when you are back in the classroom. If you are using three-ring binders, do not bring them outdoors. Take the paper out, put it on a clipboard, and return it to the binder when back in the classroom. Don't take the risk of the rings opening and everything blowing away.

Elastic bands around the bottom of the clipboard, or around a stiffer composition notebook, will help keep the paper from flapping around and becoming too weathered.

Bring Writing Tools Outdoors

You can bring chart paper outdoors. Roll up a blank piece of chart paper, grab some blue painter's tape, and stick the piece of paper to the school wall. You'll need to tape all four corners. Or set up a chart inside, and clip it to a chain-link fence with binder clips or clothespins when you go outdoors.

Take along extra pencils, as pencil points will break. Some teachers find it helpful to tie pencils onto clipboards. Pencils should be tucked between the clip and the notebook sheet so that students don't poke themselves or, more likely, accidentally break the pencil points.

Decide When to Write Outdoors

In general, notebook entries will be more detailed and more insightful if students can stay outdoors where the scientific exploration occurred. Sometimes, you will want to complete notebook entries indoors. If you are teaching the module early in the year when students are building up the routines for using the schoolyard, or if the weather is not ideal (a little chilly, raining, too hot, too windy), then you may want students to make notebook entries after returning to the classroom. If students are totally focused and in the moment, they can stay outdoors while they answer the focus question. If other students are outdoors playing, you may need to bring the class indoors to complete the written work. Only you will be able to determine what is best at the time.

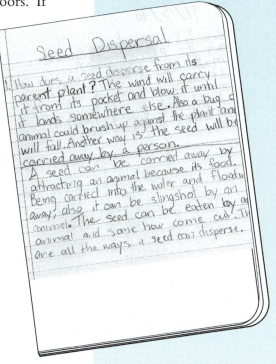

Science Notebooks in Grades 3–5

Science Notebooks in Grades 3–5

CLOSING THOUGHTS

Engaging students in active science with notebooks provides a rich experience. Doing this successfully requires thoughtful interactions among students, materials, and natural phenomena. Initially, adding notebooks to your science teaching will require you to focus students' attention on how to set up the notebook, what types of entries students should make, and when students should be using their notebooks. You will establish conventions about where to record the date and title, where to keep notebooks, how to glue notebook sheets into notebooks, and when to record observations and thinking.

Once you are past these perfunctory issues, you can shift your focus to the amount of scaffolding to provide to students or to encouraging students to create their own notebook entries. During this time, you and your students are developing skills to improve the quality of notebook entries. These skills may include asking better questions to focus students' attention on a specific part of an organism or using color to enhance a drawing. Students begin to make entries with less prompting. They give more thought to supporting their responses to the focus question. When asked to make a derivative product, students thumb through their notebooks to find the needed information. The notebook becomes a tool for students to help recall their learning.

▶ **NOTE**
For more on derivative products, see the Science-Centered Language Development chapter.

As students begin to document their thinking about focus questions and other queries, you may begin to wonder, "Should I be doing something with their notebooks?" This is when your focus shifts from the notebook as just something students use during science learning to the notebook as an assessment tool. Once everyone is comfortable recording the focus question and collecting data, you can take the next step of collecting notebooks and reading students' responses as a measure of not just how individual students are learning, but what the pervasive needs of students are. You choose next-step strategies that address students' needs before proceeding to the next investigation. The notebooks act as an assessment tool that lets you modify your science instruction.

This process will take time, discussions with colleagues, revisiting different sections of this chapter, and critical scrutiny of students' work before both you and your students are using notebooks to their full potential.

Science-Centered Language Development

Science-Centered Language Development

Teams of science inquirers talk about and write about their questions, their tentative explanations, their relationships between evidence and explanations, and their reasons and judgments about public presentations and scientific arguments in behalf of their work. It is in the context of this kind of scientific activity that students' literacy of the spoken and written word develops along with the literacy of the phenomenon.

Hubert M. Dyasi, "Visions of Inquiry: Science"

Contents

Introduction 1
The Role of Language in Scientific and Engineering Practices 3
Speaking and Listening Domain 6
Writing Domain 12
Reading Domain 21
Science-Vocabulary Development 30
English-Language Development 36
References 42

INTRODUCTION

In this chapter, we explore the intersection of science and language and the implications for effective science teaching and language development. We identify best practices in language arts instruction that support science learning and examine how learning science content and practices supports language development. The active investigations, science notebooks, *FOSS Science Resources* readings, and formative assessment activities in FOSS provide rich contexts in which students develop and exercise thinking processes and communication skills. Together, these elements comprise effective instruction in both science and language arts—students experience the natural world around them in real and authentic ways and use language to inquire, process information, and communicate their thinking about scientific phenomena. We refer to the development of language within the context of science as science-centered language development.

Science-Centered Language Development

Language plays two crucial roles in science learning: (1) it facilitates the communication of conceptual and procedural knowledge, questions, and propositions (external, public), and (2) it mediates thinking, a process necessary for understanding (internal, private). These are also the ways scientists use language: to communicate with one another about their inquiries, procedures, and understandings; to transform their observations into ideas; and to create meaning and new ideas from their work and the work of others. For students, language development is intimately involved in their learning about the natural world. Science provides a real and engaging context for developing literacy, and language arts skills and strategies support conceptual development and scientific practices. For example, the skills and strategies used for reading comprehension, writing expository text, and oral discourse are applied when students are recording their observations, making sense of science content, and communicating their ideas. Students' use of language improves when they discuss, write, and read about the concepts explored in each investigation.

We begin our exploration of science and language by focusing on language functions and how specific language functions are used in science to facilitate information acquisition and processing (thinking). Then we address issues related to the specific language domains—speaking and listening, writing, and reading. Each section addresses

- how skills in that domain are developed and exercised in FOSS science investigations;
- literacy strategies that are integrated purposefully into the FOSS investigations;
- suggestions for additional literacy strategies that both enhance student learning in science and develop or exercise English-language literacy skills.

Following the domain discussions is a section on science-vocabulary development, with scaffolding strategies for supporting all learners. The last section covers language-development strategies specifically for English learners.

▶ **NOTE**
The term *English learners* refers to students who are learning to understand English. This includes students who speak English as a second language and native English speakers who need additional support to use language effectively.

THE ROLE OF LANGUAGE IN SCIENTIFIC AND ENGINEERING PRACTICES

Language functions are the purpose for which speech or writing is used and involve both vocabulary and grammatical structure. Understanding and using language functions appropriately is important in effective communication. Students use numerous language functions in all disciplines to mediate communication and facilitate thinking (e.g., they plan, compare, discuss, apply, design, draw, and provide evidence).

In science, language functions facilitate the enactment of scientific and engineering practices. For example, when students are *collecting data*, they are using language functions to identify, label, enumerate, compare, estimate, and measure. When students are *constructing explanations*, they are using language functions to analyze, communicate, discuss, evaluate, and justify.

A Framework for K–12 Science Education: Practices, Crosscutting Concepts, and Core Ideas (National Research Council 2012) provides a "set of essential practices as essential for any education in the sciences and engineering." Each of these scientific and engineering practices requires the use of multiple language functions. Often, these language functions are part of an internal dialogue weighing the merits of various explanations—what we call thinking. The more language functions with which we are facile, the more effective and creative our thinking can be.

The scientific and engineering practices are listed below, along with a sample of the language functions that are exercised when effectively engaged in that practice. (Practices are bold; language functions are italic.)

Asking questions and defining problems

- *Ask* questions about objects, organisms, systems, and events in the natural world (science) or define a problem (engineering).

Planning and carrying out investigations; analyzing, and interpreting data

- *Plan* and conduct investigations.
- *Observe* and *record* data.
- *Measure* to extend the senses to acquire data.
- *Organize* observations (data), using numbers, words, images, and graphics.

Examples of Language Functions
Analyze
Apply
Ask questions
Clarify
Classify
Communicate
Compare
Conclude
Construct
Critique
Describe
Design
Develop
Discuss
Distinguish
Draw
Enumerate
Estimate
Evaluate
Experiment
Explain
Formulate
Generalize
Group
Identify
Infer
Interpret
Justify
Label
List
Make a claim
Measure
Model
Observe
Organize
Plan
Predict
Provide evidence
Reason
Record
Represent
Revise
Sequence
Solve
Sort
Strategize
Summarize

Science-Centered Language Development

Constructing explanations (science) and designing solutions (engineering); engaging in argument from evidence

- *Predict* future events, based on evidence and reasonings.
- Use data and logic to *construct* and *communicate* reasonable explanations.
- *Develop*, *discuss*, *evaluate*, and *justify* the merits of explanations.
- Use science-related technologies to *apply* scientific knowledge and *solve* problems.

Developing and using models

- Create models to *explain* natural phenomena and *predict* future events or outcomes.
- *Describe* how models represent the natural world.
- *Compare* models to actual phenomena and *identify* limitations of the models.

Obtaining, evaluating, and communicating

- *Identify* key ideas in text, identify supporting evidence, and know how to use various text features.
- *Distinguish* opinion from evidence.
- Use writing as a tool for *clarifying* ideas and *communicating*.

Research supports the claim that when students are intentionally using language functions in thinking about and communicating in science, they improve not only science content knowledge, but also language-arts and mathematics skills (Ostlund 1998; Lieberman and Hoody 1998). Language functions play a central role in science as a key cognitive tool for developing higher-order thinking and problem-solving abilities that, in turn, support academic literacy in all subject areas.

Here is an example of how an experienced teacher can provide an opportunity for students to exercise language functions in FOSS. In the **Soils, Rocks, and Landforms Module**, this is one piece of content we expect students to have acquired by the end of the module.

- The greater the flow of water across Earth's surface, the greater the rate of erosion and deposition.

The scientific practices the teacher wants the class to focus on are *interpreting data* and *constructing explanations*.

The language functions students will exercise while engaging in these scientific practices are *comparing*, *explaining*, and *providing evidence*. The teacher understands that these language functions are appropriate to the purpose of the science investigation and support the Common Core Standards for writing (students will write explanatory texts to examine a topic and convey ideas and information clearly).

- Students will *compare* observational data (from a stream-table investigation) to *explain* the relationship between the amount of water that runs over a surface and the amount of erosion and deposition that occur.

The teacher can support the use of language functions by providing structures such as sentence frames.

- As _____, then _____.

    As more water flows over a surface, then more erosion and deposition occur.

- When I changed _____, then _____ happened.

    When I changed the amount of water flowing down the stream table, then more erosion and deposition happened.

- The more/less _____, the _____.

    The more water that flows across the earth materials, the more erosion and deposition that occur.

> **NOTE**
> For more examples of how FOSS teachers address language-arts standards while conducting science investigations with students, go to FOSSweb (www.FOSSweb.com).

Science-Centered Language Development

Science-Centered Language Development

SPEAKING AND LISTENING DOMAIN

The FOSS investigations are designed to engage students in productive oral discourse. Talking requires students to process and organize what they are learning. Listening to and evaluating peers' ideas calls on students to apply their knowledge and to sharpen their reasoning skills. Guiding students in small-group and whole-class discussions is critical to the development of conceptual understanding of the science content and the ability to think and reason scientifically.

FOSS investigations start with a discussion—either a review to activate prior knowledge or presentation of a focus question or a challenge to motivate and engage active thinking. During active investigation, students talk with one another in small groups, share their observations and discoveries, point out connections, ask questions, and start to build explanations. The discussion icon in the sidebar of the *Investigations Guide* indicates when small-group discussions should take place.

During the activity, the *Investigations Guide* indicates where it is appropriate to pause for whole-class discussions to guide conceptual understanding. The *Investigations Guide* provides you with discussion questions to help stimulate student thinking and support sense making. At times, it may be beneficial to use sentence frames or standard prompts to scaffold the use of effective language functions and structures.

At the end of the investigation, there is another opportunity to develop language through speaking and listening in the Wrap-Up/Warm-Up. During this time, students are often asked to discuss their responses to the focus question. All students need to be engaged in this opportunity to use different language functions.

On the following pages are some suggestions for providing structure to those discussions and for scaffolding productive discourse when needed. Teaching techniques used to generate discussion in language arts and other content areas can also be used effectively during science. Using the protocols that follow will ensure inclusion of all students in discussions.

Partner and Small-Group Discussion Protocols

Whenever possible, give students time to talk with a partner or in a small group before conducting a whole-class discussion. This provides all students with a chance to formulate their thinking, express their ideas, practice using the appropriate science vocabulary, and receive input from peers. Listening to others communicate different ways of thinking about the same information from a variety of perspectives helps students negotiate the difficult path of sense making for themselves.

Dyads. Students pair up and take turns either answering a question or expressing an idea. Each student has 1 minute to talk while the other student listens. While student A is talking, student B practices attentive listening. Student B makes eye contact with student A, but cannot respond verbally. After 1 minute, the roles reverse.

Here's an example from the **Mixtures and Solutions Module**. Just before students answer the focus question in their notebooks, you ask students to pair up and take turns sharing their answer to the question "How do you know when a chemical reaction has occurred?" The language objective is for students to be able to explain what happens during a chemical reaction and provide evidence (orally and in writing) that it has occurred. These sentence frames can be written on the board to scaffold student thinking and conversation.

- I claim that _____ produces a chemical reaction, because _____.

- Evidence for my claim includes _____.

Partner parade. Students form two lines facing each other. Present a question, an idea, an object, or an image as a prompt for students to discuss. Give students 1 minute to greet the person in front of them and discuss the prompt. After 1 minute, call time. Have the first student in one of the lines move to the end of the line, and have the rest of the students in that line shift one step sideways so that everyone has a new partner. (Students in the other line do not move.) Give students a new prompt to discuss for 1 minute with their new partners.

For example, students are just beginning the investigation on landforms, and you want to assess prior knowledge. Give each student a picture of a landform, and have students line up facing each other. For the first round, ask, "What do you know about the image on your card?" For the second round, ask, "How do you think the landform in your picture formed?" For the third round, ask, "What personal experience have you had with this landform?" The language objective is for students to describe their observations, infer how the landform formed,

Partner and Small-Group Discussion Protocols
- *Dyads*
- *Partner parade*
- *Put in your two cents*

Science-Centered Language Development

Science-Centered Language Development

and reflect upon and relate any experiences they may have had with a similar landform. The following sentence frames can be used to scaffold student discussion.

- I notice _____.
- I think the landform was formed by _____.
- The landform in my picture reminds me of the time when _____.

Put in your two cents. For small-group discussions, give each student two pennies or similar objects to use as talking tokens. Each student takes a turn putting a penny in the center of the table and sharing his or her idea. Once all have shared, each student takes a turn putting in the other penny and responding to what others in the group have said. For example,

- I agree (or don't agree) with _____ because _____.

Here's an example from the **Sun, Moon, and Planets Module**. Students have been monitoring their shadows throughout the day and are still struggling to make coherent connections between Earth's rotation and the effect that it has on shadows. The language objective is for students to describe their observations, explain why the direction and length of their shadows change during the day, and provide evidence based on their own observations. You give each student two pennies, and in groups of four, they take turns putting in their two cents. For the first round, each student answers the question "Why do shadows change during the day?" They use the frame

- I think shadows change during the day because _____.
- My evidence is _____ .

On the second round, each student states whether he or she agrees or disagrees with someone else in the group and why, using the sentence frame.

Whole-Class Discussion Supports

The whole-class discussion is a critical part of sense making. After students have had the active learning experience and have talked with their peers in partners and/or small groups, sharing their observations with the whole class sets the stage for developing conventional explanatory models. Discrepant events, differing results, and other surprises are discussed, analyzed, and resolved. It is important that students realize that science is a process of finding out about the world around them. This is done through asking questions, testing ideas, forming explanations, and subjecting those explanations to logical

Whole-Class Discussion Supports
- *Sentence frames*
- *Guiding questions*

scrutiny. Leading students through productive discussion helps them connect their observations and the abstract symbols (words) that represent and explain those observations. Whole-class discussion also provides an opportunity for you to interject an accurate and precise verbal summary statement as a model of the kind of thinking you are seeking. Facilitating effective whole-class discussions takes skill, practice, a shared set of norms, and patience. In the long run, students will have a better grasp of the content and improve their abilities to think independently and communicate effectively.

Norms should be established so that students know what is expected during science discussions.

- Science content and practices are the focus.
- Everyone participates.
- Ideas and experiences are shared, accepted, and valued. Everyone is respectful of one another.
- Claims are supported by evidence.
- Challenges (debate and argument) are part of the quest for complete understanding.

The same discussion techniques used during literacy circles and other whole-class discussions can be used during science instruction (e.g., attentive listening, staying focused on the speaker, asking questions, responding appropriately). In addition, in order for students to develop and practice their reasoning skills, they need to know the language forms and structures and the behaviors used in evidence-based debate and argument (e.g., using data to support claims, disagreeing respectfully, asking probing questions; Winokur and Worth 2006).

Explicitly model and conduct mini-lessons (5 to 10 minutes of focused instruction) on the language structures appropriate for active discussions, and provide time for students to practice them, using the science content.

Sentence frames. The following samples can be posted as a scaffold as students learn and practice their reasoning and oral participation skills.

- I think _____, because _____.
- I predict _____, because _____.
- I claim _____; my evidence is _____.
- I agree with _____ that _____.
- My idea is similar/related to _____'s idea.

> **TEACHING NOTE**
>
> Let students know that scientists change their minds based on new evidence. It is expected that students will revise their thinking based on evidence presented in discussions.

> **TEACHING NOTE**
>
> Encourage science talk. Allow time for students to engage in discussions that build on other students' observations and reasoning. After an investigation, use a teacher- or student-generated question, and either just listen or facilitate the interaction with questions to encourage expression of ideas among students.

Science-Centered Language Development

Science-Centered Language Development

- I learned/discovered/heard that ____.
- <Name> explained ____ to me.
- <Name> shared ____ with me.
- We decided/agreed that ____.
- Our group sees it differently, because ____.
- We have different observations/results. Some of us found that ____. One group member thinks that ____.
- We had a different approach/idea/solution/answer ____.

Guiding questions. The *Investigations Guide* provides questions to help concentrate student thinking on the concepts introduced in the investigation. Guiding questions should be used during the whole-class discussion to facilitate sense making. Here are some other open-ended questions that help guide student thinking and promote discussion.

- What did you notice when ____?
- What do you think will happen if ____?
- How might you explain ____? What is your evidence?
- What connections can you make between ____ and ____?

Whole-Class Discussion Protocols

The following examples of tried-and-true participation protocols can be used to enhance whole-class discussions during science and all other curriculum areas. The purpose of these protocols is to increase meaningful participation by giving all students access to the discussion, allowing students time to think (process), and providing a context for motivation and engagement.

Think-pair-share. When asking for a response to a question posed to the class, allow time for students to think silently for a minute. Then, have students pair up with a partner to exchange thoughts before you call on a student to share his or her ideas with the whole class.

Pick a stick. Write each student's name on a craft stick, and keep the sticks handy at the front of the room. When asking for responses, randomly pick a stick, and call on that student to start the discussion. Continue to select sticks as you continue the discussion. Your name can also be on a stick in the cup. You can put the selected sticks in a different location or back into the same cup to be selected again.

Whip around. Each student takes a quick turn sharing a thought or reaction. Questions are phrased to elicit quick responses that can be expressed in one to five words (e.g., "Give an example of a stored-energy source." "What does the word *heat* make you think of?").

Group posters. Have small groups design and graphically record their investigation data and conclusions on a quickly generated poster to share with the whole class.

Whole-Class Discussion Protocols
- *Think-pair-share*
- *Pick a stick*
- *Whip around*
- *Group posters*

Two-cup pick-a-stick container

One-cup pick-a-stick container divided with tape

Science-Centered Language Development

Science-Centered Language Development

WRITING DOMAIN

Information processing is enhanced when students engage in informal writing. When allowed to write expressively without fear of being scorned for incorrect spelling or grammar, students are more apt to organize and express their thoughts in different ways that support sense making. Writing promotes the use of scientific practices, thereby developing a deeper engagement with the science content. It also provides guidance for more formal derivative science writing (Keys 1999).

Science Notebooks

The science notebook is an effective tool for enhancing learning in science and exercising various forms of writing. Science notebooks provide opportunities both for expressive writing (students craft explanatory narratives that make sense of their science experiences) and for practicing informal technical writing (students use organizational structures and writing conventions). Starting as emergent writers in kindergarten, students learn to communicate their thinking in an organized fashion while engaging in the cognitive processes required to develop concepts and build explanations. Having this developmental record of learning also provides an authentic means for assessing students' progress in both scientific thinking and communication skills.

One way to help students develop the writing skills necessary for productive notebook entries is to focus on the corresponding language functions. The language forms and structures used to perform these language functions in science are used in all curricular areas and, therefore, can be suitably taught in conjunction with existing language-arts instruction. This can be done through mini-lessons on the writing skills that support the various types of notebook entries.

Table 1, at the end of the Writing Domain section, provides examples of how language functions are used to help students develop both their general writing skills and their thinking abilities within the format of the science-notebook entry. The writing objectives for a mini-lesson, along with the particular language forms and structures (vocabulary, syntax, linking words, organization of ideas, and so on), are identified in the table along with the suggested sentence frames for scaffolding.

For example, if students are observing a particular insect's behavior, the language objective might be "Students describe in detail the behavior of the insect." A prior mini-lesson on using sensory details to describe observations would provide students with the language forms and structures appropriate for recording their data in their notebooks during the observations. As a scaffold, students could also be provided with sentence frames to help them write detailed narratives.

> **NOTE**
> For more information about supporting science-notebook development, see the Science Notebooks chapter.

> **NOTE**
> Language forms and structures refer to the internal grammatical structure of words and how those words go together to make sentences.

Language function	Language objectives for writing in notebooks	Language forms, structures, and scaffolds for writing
Notebook component—data acquisition		
Describe	Write narratives: use details, sensory observations, and connections to prior knowledge.	I observed ___. When I touch the ___, I feel ___. The ___ has ___. I noticed ___. It feels ___. It smells ___. It sounds ___. It reminds me of ___, because ___.

This sample from Table 1 shows how language functions can be developed and applied when writing in science notebooks.

> **NOTE**
> The complete table appears at the end of this Writing Domain section.

Developing Derivative Language-Arts Products

Science notebooks provide students with a source of information (content) from which they can draw to create more formal science-centered writing pieces. Derivative products are written pieces that are generated with specific language-arts goals in mind, such as audience and language function. Writing-to-learn methods can enhance science concepts when students engage in different types of writing for different purposes (Hand and Prain 2002). We know that students are more engaged and motivated to write when they have a clear and authentic context for writing.

The language extensions in the Interdisciplinary Extensions section at the end of each investigation suggest writing activities that can be used to help students learn the science content for that particular investigation. The writing activities incorporate language-arts skills appropriate for the grade level. Questions and ideas for future writing activities that surface during the investigations can be recorded in a class list, in science notebooks, or in students' writing folders. Here are general suggestions for using science content to create products in each of several writing genres.

Descriptive writing. Students use descriptive writing to portray an organism, an environment, an object, or a phenomenon in such a way that the reader can almost recapture the writer's experience. This is done through the use of sensory language; rich, vivid, and lively detail; and figurative language, such as simile, hyperbole, metaphor, and symbolism. You can remind students to show, rather than tell, through the use of active verbs and precise modifiers.

To help them learn the science content, students can use the information in their science notebooks to elaborate on their observations by using descriptive vocabulary, analogies, and drawing.

Derivative Language-Arts Products
- *Descriptive writing*
- *Persuasive writing*
- *Narrative writing*
- *Expository writing*
- *Recursive cycle*

Science-Centered Language Development

For example, kindergartners can draw pictures of isopods, fish, worms, things made of wood, different types of trees and leaves, and so on. Through drawings and words, students describe objects' properties and compare how those objects are the same and different.

Primary students can make property cards by writing on an index card as many properties as they can that describe an object or organism. Then, students take turns reading the properties to another student to see if the partner can identify the corresponding object.

Students of all ages can expand on their notebook entries in poetry that expresses their interpretation of organisms, objects, and phenomena such as crayfish, seeds, electricity, rivers, weather, and minerals. The use of similes is a good way to engage students in making comparisons. You can use a simple frame, such as: _____ is like _____. (For example, in the **Air and Weather Module**, students might write "The weather today is like a dragon.")

Students can share their similes. Other students can explain why they think the simile works.

Persuasive writing. The objective of persuasive writing is to convince the reader that a stated opinion or interpretation of data is worthwhile and meaningful. Students learn to support their claims with evidence and to use persuasive techniques, such as logical arguments and calls to action. By using claims and evidence to formulate conclusions in their science notebooks, students develop and apply their thinking and reasoning skills to form the basis of persuasive writing in a variety of formats, such as essays, letters, editorials, advertisements, award nominations, informational pamphlets, and petitions. Animal habitats, energy use, weather patterns, landforms, and water sources are just a few science topics that can generate questions and issues for persuasive writing.

Here is a sample of persuasive writing frames (modified from Gibbons 2002).

> Title: _____
> The topic of this discussion is _____.
> My opinion (position, conclusion) is _____.
> There are <number> reasons why I believe this to be true.
> First, _____.
> Second, _____.
> Finally, _____.
> On the other hand, some people think _____.
> I have also heard people say _____.
> However, my claim is that _____ because _____.

Narrative writing. Science provides a broad landscape of engaging material for stimulating the imagination for the writing of stories, songs, biographies, autobiographies, poems, and plays. Students can use organisms or objects as characters; describe habitats and environments as settings; and write scripts portraying various systems, such as weather patterns, flow of electricity, and water, rock, or life cycles.

Expository writing. Students use science content to inform, explain, clarify, define, or instruct through writing letters, definitions, procedures, newspaper and magazine articles, posters, pamphlets, and research reports. Expository writing is characterized by a focus on main topics with supporting facts, details, explanations, and examples and is organized in a clear, coherent, and sequential manner. Expository writing has a clear focus on communicating accurate, complete, and detailed representations of science content and/or observation of natural history or events.

During writing instruction, students can use the information in their science notebooks and in related readings (and other sources, such as video content) to write a more formal and conclusive answer to the focus question. Strategies such as the writing process (plan, draft, edit, revise, and share) and writing frames (modeling and guiding the use of topic sentences, transition and sequencing words, examples, explanations, and conclusions) can be used with the science content to develop proficiency in critical writing skills.

> **NOTE**
> Human characteristics should not be given to organisms (anthropomorphism) in science investigations, only in literacy extensions.

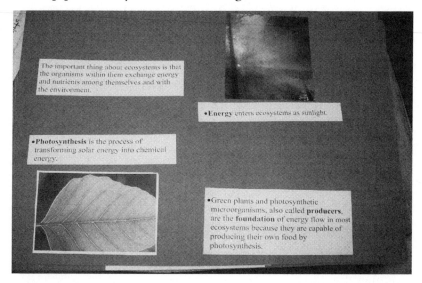

A page from a big book about ecosystems, a derivative product created during language arts

Science-Centered Language Development

Science-Centered Language Development

Here are three samples of writing frames for science reports for primary and upper-elementary students (modified from Wellington and Osborne 2001).

Primary (life cycle appropriate for the **Insects and Plants Module**)

Title: _____

(Define) A _____ is _____.

(Classify) A _____ is a kind of _____.

(Describe) A _____ is _____.

(Habitat) _____ lives _____.

(Life cycle) _____ started life as _____.

(Change over time) Then it _____.

(Food) A _____ eats _____.

(Evidence) I know that _____ because _____.

Upper elementary (living structures appropriate for the **Structures of Life Module**)

Title: _____

(Identify) The part of the body I am describing is the _____.

(Describe) It consists of _____.

(Explain) The function of these parts is _____.

(Example) This drawing shows _____.

Upper elementary (explanation)

Title: _____

I want to explain why (how) _____.

An important reason for why (how) this happens is that _____.

Another reason is that _____.

I know this because _____.

Recursive cycle. An effective method for extending students' science learning through writing is the recursive cycle of research (Bereiter 2002). This strategy emphasizes writing as a process for learning, similar to the way students learn during the active science investigations.

1. Decide on a problem or question to write about.
2. Formulate an idea or a conjecture about the problem or question.
3. Identify a remedy or an answer, and develop a coherent discussion.
4. Gather information (from science notebooks, *FOSS Science Resources*, books, FOSSweb interactives, Internet, interviews, videos, experiments, etc.).
5. Reevaluate the problem or question based on what has been learned.
6. Revise the idea or conjecture.
7. Make presentations (reports, posters, electronic presentations, etc.).
8. Identify new needs, and make new plans.

This process can continue for as long as new ideas and questions occur, or students can present a final product in any of the suggested formats.

Science-Centered Language Development

Table 1. Examples of how language functions are exercised in science notebook writing and examples of sentence frames and language structures students may use

Language function	Language objectives for writing in notebooks	Language forms, structures, and scaffolds for writing
Organizational setup		
Organize	Set up and organize notebook: table of contents, glossary or index, page numbers, date, turning to the next blank page.	Use structures such as rows, columns, blocks, numbers, location, alphabetizing.
Notebook component—planning the investigation		
Design Strategize	Write a narrative plan: communicate ideas on an approach to answer the focus question or challenge posed in the investigation.	Use adverbs such as *first*, *second*, *next*, *then*, *finally*. Brainstorming ideas: First I will ____, and then I will ____. I will need to ____ to ____.
List Record	Write ordered lists: materials, variables, vocabulary words, bullets	We need ____, ____, ____, and ____ to ____ .
Sequence	Record step-by-step procedures: explain exactly what to do, number steps in a sequence, specifics for measurement, how controls will be established, how variables will be measured (tools/units).	1. _____. 2. _____. 3. _____. I will change ____. I will not change ____. I will measure ____. I will not measure ____.
Notebook component—data acquisition		
Describe	Write narratives: use details, sensory observations, connections to prior knowledge.	I observed ____. When I touch the ____, I feel ____. The ____ has ____. I noticed ____. It feels ____. It smells ____. It sounds ____. It reminds me of ____, because ____.

Table 1 (*continued*)

Language function	Language objectives for writing in notebooks	Language forms, structures, and scaffolds for writing
Notebook component—data acquisition (continued)		
Draw Label Identify	Make technical drawings: draw large, accurate, and detailed representations; identify parts of a system.	Label drawing, using science vocabulary. Recognize shapes, form, location, color, size, and scale. My drawing shows _____.
Organize Compare Classify	Make charts and tables: use a T-table or chart for recording and displaying data.	Set up rows, columns, headings. My T-table compares _____.
Sequence Compare	Record changes: use language structures to communicate change over time, cause and effect.	At first, _____, but now _____. We saw that first _____, then _____, and finally _____. When I _____, it _____. After I _____, it _____.
Notebook component—data organization		
Enumerate	Plot graphs: decide when and how to use bar graphs, line plots, and two-coordinate graphs to organize data.	Define x- and y-axes; provide axis labels, title, units, coordinates; use equal intervals, set origin = (0, 0).
Compare Classify Sequence	Use graphic organizers and narratives to express similarities and differences, to assign an object or action to the category or type to which it belongs, and to show sequencing and order.	This _____ is the same as _____ because _____. This _____ is different than _____ because _____. All these are _____ because _____. _____, _____, and _____ all have/are _____.
Analyze	Use graphic organizers, narratives or concept maps to identify part/whole or cause-and-effect relationships.	Use relationship verbs such as *contain*, *consist of*. As _____, then _____. When I changed _____, then _____ happened. The more/less _____, then _____.

Science-Centered Language Development

Science-Centered Language Development

Table 1 (*continued*)

Language function	Language objectives for writing in notebooks	Language forms, structures, and scaffolds for writing
Notebook component—making sense of data		
Infer Explain	Provide claims and evidence: write assertions about what was learned from the investigation, use the data as evidence to support those claims.	Use inferential logical connectors such as *although*, *while*, *thus*, *therefore*. I claim that ____. I know this because ____.
Provide evidence	Use qualitative and quantitative data from the investigation as evidence to support claims.	Use qualitative descriptors such as *more/less*, *longer/shorter*, *brighter/dimmer*. Use quantitative expressions using standard metric units of measurement such as cm, mL, °C. My data show ____.
Summarize Predict Generalize	Write a summary narrative to communicate what was learned; ask questions and make predictions based on the newly acquired knowledge.	Answer the focus question by rewriting it as a statement and providing evidence from data. Make a concluding statement. I learned ____. Therefore, I think ____. I predict ____ because ____. A new question I have is ____.
Notebook component—next-step strategies		
Critique Evaluate	Reflect on experience: review notebook entries and revise, using line of learning or 3 Cs (correct in red, confirm in green, and complete in blue).	I used to think ____, but now I think ____. I have changed my thinking about ____. I am confused about ____ because ____. I wonder ____.

READING DOMAIN

Reading is an integral part of science learning. Just as scientists spend a significant amount of their time reading one another's published works, students need to learn to read scientific text—to read effectively for understanding with a critical focus on the ideas being presented.

The articles in *FOSS Science Resources* help facilitate sense making as students make connections to the science concepts introduced and explored during the active investigations. Concept development is most effective when students are allowed to experience organisms, objects, and phenomena firsthand before engaging the concepts in text. The text and illustrations help students make connections between what they have experienced concretely and the abstract ideas that explain their observations.

FOSS Science Resources supports developing literacy skills by providing reading material that corresponds exactly to the concrete, personal experience provided in the active investigations. Students read with enthusiasm when they recognize familiar materials, organisms, and activities and are eager to tackle the reading to confirm their prior knowledge and discover more about the topic. In addition to making connections, once engaged, students naturally use other reading-comprehension strategies, such as asking questions, visualizing, inferring, and synthesizing, to help them understand the reading. As students apply these strategies, they are, in effect, using some of the same scientific thinking processes that promote critical thinking and problem solving.

Reading in the Primary Grades

In the kindergarten modules, you can enhance science learning by using trade books and other read-aloud resources to engage students and provide topics for lively discussions. Reading aloud helps primary students understand the science content and lets you model reading-comprehension strategies, such as asking yourself questions (thinking aloud) and summarizing a paragraph just read. The Reading in *FOSS Science Resources* sections (in every part that has a reading) usually offer suggestions for activating prior knowledge before reading, indicate places to pause and discuss key points during the reading, and describe activities to deepen understanding after the reading. As students develop their reading skills, you might try these different ways to read from *FOSS Science Resources*.

Science-Centered Language Development

Science-Centered Language Development

- Read aloud from the big book while students follow along in their own books.
- Lead students in small guided reading groups.
- Have students read aloud with a partner.
- Have students read silently on their own.

The same strategies used in language arts can be applied to reading in science. Begin with reading the article aloud so that students can hear the content read fluently and listen for meaning and coherence. Go back and review the questions within the article. Use think-pair-share or other discussion protocols to allow students to think first, share with a partner, and then respond to the group. Emphasize blending phonemes as you read the text again, and model tracking, connecting spoken words with written words, and helping students identify sight words. Model reading-comprehension strategies by using think-alouds (as you think aloud, you explain the process that you are using in order to understand the text while reading). The expository text structure also provides the opportunity for primary students to learn how to extract information from a table of contents, a glossary, an index, and other text conventions such as headings, subheads, boldface and italic print, labeled graphics, and captions.

At the end of articles, use the questions provided to guide understanding and to assess comprehension and vocabulary acquisition. Choose one or two questions for students to answer in their notebooks. Emphasize the importance of science vocabulary and the appropriate language forms and structures. Here are some ways students can enhance reading comprehension.

- Have students predict the sequence of events or content.
- Have students write or dictate questions about the text and illustrations.
- Use visualization with students to "see, touch, feel, smell, hear" in their minds the content presented in the article.
- Ask students to make connections to information in the article and their observations during the active investigation.

Reading in the Upper-Elementary Grades

As students progress from *learning to read* to *reading to learn*, they apply the strategies and skills of reading comprehension to learning about science in *FOSS Science Resources* and other texts. Writing becomes increasingly important in helping students make sense of the readings as well as the active investigations. Use the suggested questions in the *Investigations Guide* to support comprehension as students read from *FOSS Science Resources*. For most of the investigation parts, the articles are designed to follow the active investigation and are interspersed throughout the flow of the module. This allows students to acquire the necessary background knowledge through active experience before tackling the wider-ranging content and relationships presented in the text. Additional strategies for reading are derived from the seven essential strategies that readers use to help them understand what they read (Keene and Zimmermann 2007).

- Monitor for meaning: Discover when you know and when you don't know.

- Use and create schemata: Make connections between the novel and the known; activate and apply background knowledge.

- Ask questions: Generate questions before, during, and after reading that reach for deeper engagement with the text.

- Determine importance: Decide what matters most, what is worth remembering.

- Infer: Combine background knowledge with information from the text to predict, conclude, make judgments, and interpret.

- Use sensory and emotional images: Create mental images to deepen and stretch meaning.

- Synthesize: Create an evolution of meaning by combining understanding with knowledge from other texts/sources.

Following are some strategies that enhance the reading of expository texts in general and have proven to be particularly helpful in science.

Build on background knowledge. Activating prior knowledge is critical for helping students make connections between what they already know and new information. Reading comprehension improves when students have the opportunity to think, discuss, and write about what they know about a topic before reading. Review what students learned from the active investigation, provide prompts for making connections, and ask questions to help students recall past experiences and previous exposure to concepts related to the reading.

Strategies for Reading in Upper-Elementary Grades

- *Build on background knowledge*
- *Create an anticipation guide*
- *Draw attention to vocabulary*
- *Preview the text*
- *Turn and talk*
- *Jigsaw text reading*
- *Note making*
- *Interactive reading aloud*
- *Summarize and synthesize*
- *3-2-1*
- *Write reflections*
- *Preview and predict*
- *SQ3R*

Science-Centered Language Development

Science-Centered Language Development

Create an anticipation guide. Create true-or-false statements related to the key ideas in the reading selection. Ask students to indicate if they agree or disagree with each statement before reading, then have them read the text, looking for the information that supports their true-or-false claims. Anticipation guides connect students to prior knowledge, engage them with the topic, and encourage them to explore their own thinking.

Draw attention to vocabulary. Check the article for bold words to determine if there are words students may not know. Review the science words that have already been listed on the word wall and in students' notebooks. For new science and nonscience vocabulary words that appear in the reading, have students predict their meanings before reading. During the reading, have students use strategies such as context clues and word structure to see if their predictions were correct. This strategy activates prior knowledge and engages students by encouraging analytical participation with the text.

Preview the text. Give students time to skim through the selection, noting subheads, before reading thoroughly. Point out the particular structure of the text and what discourse markers to look for. For example, most *FOSS Science Resources* articles are written as cause and effect, problem and solution, question and answer, comparison and contrast, description, and sequence. Students will have an easier time making sense of the text if they know what to look for in terms of text structure. Model and have students practice analyzing these different types of expository text structures by looking for examples, patterns, and the discourse markers.

Point out how *FOSS Science Resources* text is organized (titles, headings, subheadings, questions, and summaries) and how to use the table of contents, glossary, and index. Explain how to scan for formatting features that provide key information (such as bold type and italics, captions, and framed text) and graphic features (such as tables, graphs, photographs, maps, diagrams, charts, and illustrations) that help clarify, elaborate, and explain important information in the reading.

While students preview the article, have them focus on the questions that appear in the text, as well as questions at the end of the article. Encourage students to write down questions they have that they think the reading will answer.

> ▶ **NOTE**
> Discourse markers are words or phrases that relate one idea to another. Examples are "however," "on the other hand," and "second."

Turn and talk. When reading as a whole class, stop at key points and have students share their thinking about the selection with the student sitting next to them or in their collaborative group. This strategy helps students process the information and allows everyone to participate in the discussion. When reading in pairs, encourage students to stop and discuss with their partners. One way to encourage engagement and understanding during paired reading is to have students take turns reading aloud a paragraph or section on a certain topic. The one who is listening then summarizes the meaning conveyed in the passage.

Jigsaw text reading. Students work together in small groups (expert teams) to develop a collective understanding of a text. Each expert team is responsible for one portion of the assigned text. The teams read and discuss their portions to gain a solid understanding of the key concepts. They might use graphic organizers to refine and organize the information. Each expert team then presents its piece to the rest of the class. Or new, small jigsaw groups can be formed that consist of at least one representative from each expert team. Each student shares with the jigsaw group what their team learned from their particular portion of the text. Together, the participants in the jigsaw group fit their individual pieces together to create a complete picture of the content in the article.

Note making. The more interactive that students make a reading, the better for their understanding. Encourage students to become active readers by asking them to make notes as they read. Studies have shown that note making—especially paraphrasing and summarizing—is one of the most effective means for understanding text (Graham and Herbert 2010; Applebee 1984).

Students can annotate the text by writing thoughts and questions on self-stick notes. Using symbols or codes can help facilitate comprehension monitoring. Here are some possible symbols (Harvey 1998).

- ★ interesting
- BK background knowledge
- ? question
- C confusing
- I important
- L learning something new
- W wondering
- S surprising

Science-Centered Language Development

Science-Centered Language Development

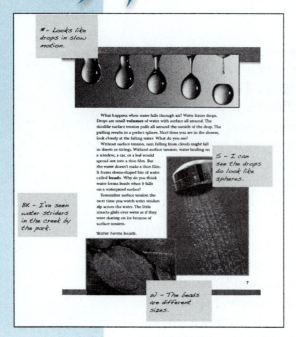

An example of annotated text in *FOSS Science Resources*

Students can use a set of symbols while making notes about connections. The readings in *FOSS Science Resources* incorporate the active learning that students gain from the investigations so that they can make authentic text-to-self (T-S) connections. In other words, what they read reminds them of firsthand experiences, making the article more engaging and easier to understand. Text-to-text (T-T) connections are notes students make when they discover a new idea that reminds them of something they've read previously in another text. Text-to-world (T-W) connections involve the text and more global everyday connections to students' lives.

You can model note-making strategies by displaying a selection of text, using a projection system, a document camera, or an interactive whiteboard. As you read the text aloud, model how to write comments on self-stick notes, and use a graphic organizer in a notebook to enhance understanding.

Graphic organizers help students focus on extracting the important information from the reading and analyzing relationships between concepts. This can be done by simply having students make columns in their notebooks to record information and their thinking (Harvey and Goudvis 2007). Here are two examples of graphic organizers.

Notes	Thinking

Facts	Questions	Responses

Interactive reading aloud. Reading aloud is an effective strategy for enhancing text comprehension. It offers opportunities to model specific reading-comprehension strategies and allows students to concentrate on making sense of the content. When modeling, share the thinking processes used to understand the reading (questioning, visualizing, comparing, inferring, summarizing, etc.), then have students share what they observed you thinking about as an active reader.

Summarize and synthesize. Model how to pick out the important parts of the reading selection. Paraphrasing is one way to summarize. Have students record summaries of the reading, using their own words. To scaffold the learning, use graphic organizers to compare and contrast, group, sequence, and show cause and effect. Another method is to have students make two columns in their notebooks. In one column, they record what is important, and in the other they record their personal responses (what the reading makes them think about). When writing summaries, tell students,

- *Pick out the important ideas.*
- *Restate the main ideas in your own words.*
- *Keep it brief.*

3-2-1. This strategy gives students the opportunity to synthesize information and formulate questions they still have regarding the concepts covered in an article. In their notebooks, students write three new things they learned, two interesting things worth remembering and sharing, and one question that occurred to them while reading the article. Other options might include three facts, two interesting ideas, and one insight about themselves as learners; three key words, two new ideas, and one thing to think about (modified from Black Hills Special Service Cooperative 2006).

Write reflections. After reading, ask students to review their notes in their notebooks to make any additions, revisions, or corrections to what they recorded during the reading. This review can be facilitated by using a line of learning. Students draw a line under their original conclusion or under their answer to a question posed at the end of an article. They then add any new information as a new narrative entry. The line of learning indicates that what follows represents a change of thinking.

Science-Centered Language Development

Preview and predict. Instruct students to independently preview the article, directing attention to the illustrations, photos, boldfaced words, captions, and anything else that draws their attention. Working with a partner, students discuss and write three things they think they will learn from the article. Have partners verbally share their list with another pair of students. The group of four can collaborate to generate one list. Teams report their ideas, and together you create a class list on chart paper.

Read the article aloud, or have students read with a partner aloud or silently. Referring to the preview/prediction list, discuss what students learned. Have them record the most important thing they learned from the reading for comparison with the predictions.

SQ3R. Survey! Question! Read! Recite! Review! provides an overall structure for before, during, and after reading. Students begin by previewing the text, looking for features that will help them make predictions about content. Based on their surveys, students develop questions to answer as they read. Students then read the selection. They recite by telling a partner what they've learned (their partner listens carefully to their recitation, making sure they haven't missed any important concepts) and answering any questions their partner might have. Finally, they review their questions and answers. There is value in asking students to make their entire SQ3R process explicit.

Use this script as an outline, using appropriate wording where necessary.

Survey! Question!

a. Before you read, survey (preview) the reading selection.
b. As you survey, think of questions that the reading might answer while you read the title, headings, and subheadings.
c. Read the questions at the end of the article.
d. Ask yourself, "What did my teacher say about this selection when it was assigned?"
e. Ask yourself, "What do I already know about this subject from experiments or discussions we've had in class?"

Read!

    a. While reading, look for answers to the questions you raised.

    b. Think about answers for the questions at the end of the article.

    c. Study graphics, such as pictures, graphs, and tables.

    d. Reread captions associated with pictures, graphs, and tables.

    e. Note all italicized and boldfaced words or phrases.

    f. Reduce your reading speed for difficult passages.

    g. Stop and reread parts that are not clear.

    h. Read only a section at a time, and recite after each section.

Recite!

    a. After you've read a section, make notes in your science notebook by writing information in your own words.

    b. Underline or highlight important notes in your science notebook.

    c. Use the method of recitation that best suits your particular learning style. Remember that the more senses you use, the more likely you are to remember what you read. Seeing, saying, hearing, and writing will all enhance learning.

    d. Ask questions aloud about what you just read, and summarize aloud, in your own words, what you read.

    e. Listen attentively as your partner recites.

Review!

Reviewing should be ongoing. With the whole class, review the entire process, calling on groups to talk to the class about what they've learned.

Struggling Readers

For students reading below grade level, the strategies listed above can be modified to support reading comprehension by integrating strategies from the Reading in the Primary Grades section, such as read-alouds and guided reading. Students can also follow along with the audio stories on FOSSweb. Breaking the reading down into smaller chunks, providing graphic organizers, and modeling reading-comprehension strategies can also help students who may be struggling with the text. For additional strategies for English learners, see the supported-reading strategy in the English-Language Development section of this chapter.

Science-Centered Language Development

SCIENCE-VOCABULARY DEVELOPMENT

Words play two critically important functions in science. First and most important, we play with ideas in our minds, using words. We present ourselves with propositions—possibilities, questions, potential relationships, implications for action, and so on. The process of sorting out these thoughts involves a lot of internal conversation, internal argument, weighing of options, and complex linguistic decisions. Once our minds are made up, communicating that decision, conclusion, or explanation in writing or through verbal discourse requires the same command of the vocabulary. Words represent intelligence; acquiring the precise vocabulary and the associated meanings is key to successful scientific thinking and communication.

The words introduced in FOSS investigations represent or relate to fundamental science concepts and should be taught in the context of the investigation. Many of the terms are abstract and are critical to developing science content knowledge and scientific and engineering practices. The goal is for students to use science vocabulary in ways that demonstrate understanding of the concepts the words represent—not to merely recite scripted definitions. The most effective science-vocabulary development strategies help students make connections to what they already know. These strategies focus on giving new words conceptual meaning through experience; distinguishing between informal, everyday language and academic language; and using the words in meaningful contexts.

Building Conceptual Meaning through Experience

In most instances, students should be presented with new words in the context of the active experience at the need-to-know point in the investigation. Words such as *flexible*, *magnetic*, *viscous*, *silt*, *erode*, and *electricity* are conceptually loaded and significantly abstract. Students will have a much better chance of understanding, assimilating, and remembering the new word (or new meaning) if they can connect it with a concrete experience.

The new-word icon appears in the sidebar when you introduce a word that is critical to understanding the concepts or scientific practices students will be learning and applying in the investigation. When you introduce a new word, students should

- Hear it: Students listen as you model the correct contextual use and pronunciation of the word.
- See it: Students see the new word written out. Add a visual reference (an illustration or a sample) next to the word if possible.
- Say it: Have primary-grade students repeat the word chorally and clap out the syllables.
- Write it: You write the word on the board, chart paper, sentence strip, or card. Students use the new words in context when they write in their notebooks.
- Act it: Demonstrate action words such as *separate*, *compare*, and *observe* (using total physical response).

Bridging Informal Language to Science Vocabulary

Students bring a wealth of language experience to the classroom. FOSS investigations are designed to tap into students' inquisitive natures and their excitement of discovery in order to encourage lively discussions as they explore materials in creative ways. There should be a lot of talking during science time! Your role is to help students connect informal language to the vocabulary used to express specific science concepts. As you circulate during active investigation, you continually model the use of science vocabulary. For example, when a student holds up a bottle of water and says, "I can see through it!" you might respond, "Yes, I see what you mean; the liquid in the bottle is transparent." Following are some strategies for validating students' conversational language while developing their familiarity with and appreciation for science vocabulary.

Word bubbles. Choose a word from the word wall that is widely used by students and that is a synonym for a science vocabulary word. Draw a circle on the board or chart paper, and write the word in the center. Draw lines out from the circled word, and make more circles. Ask students to call out more synonyms for the word, and write them in the outer circles. Introduce the target vocabulary word as yet one more synonym for the target word, a word that is used in science. Highlight the word, and model its correct usage and pronunciation. Encourage students to use it in their discussions and in their notebook entries. Introduce the science word that is its opposite, when appropriate.

Bridging Language Strategies
- *Word bubbles*
- *Word sorts*
- *Semantic webs*
- *Concept maps*
- *Cognitive content dictionaries*
- *Word associations*

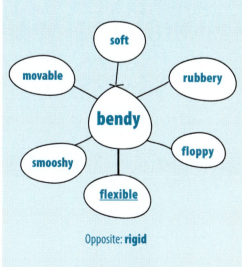

Word bubbles

Science-Centered Language Development

Science-Centered Language Development

Word sorts. Make a set of word cards from words on the word wall. Ask students to help you group the words that are synonyms or that have conceptual connections. Add the new science words to the card set. Repeat the process with a new set of words. For upper-elementary students, make sets of vocabulary cards for students to sort in small groups.

Semantic webs. Select a vocabulary word, and write it in the center of a piece of paper (or on the board if doing this with the whole class). Brainstorm a list of words or ideas that are related to the first word. Group the words and concepts into several categories, and attach them to the central word with lines, forming a web (modified from Hamilton 2002).

Concept maps. Select six to ten related science words. Have students write them on self-stick notes or cards. Have small groups discuss how the words/concepts are related. Students organize words in groups and glue them down or copy them onto a sheet of paper. Students draw lines between the related words. On the lines, they write words describing or explaining how the concept words are related.

Cognitive content dictionaries. There are many variations of frameworks to help students record new words. This example can be used with the whole class with primary students or individually with upper-elementary students to introduce a few key vocabulary words used in an investigation. Have students write the word, predict its meaning, write the final meaning after class discussion (using primary language or an illustration when appropriate), and then use the word in a sentence. The word can then be used as a signal word to call for attention.

Cognitive Content Dictionary	
New word	esker
Prediction (clues)	a flower growing at the North Pole
Final meaning (e.g., primary language, pictures)	a sand and gravel ridge, formed at the edges of a stream, flowing under a glacier
How I would use it (sentence)	The esker showed that a glacier had been in the valley at some time in the past.

Word associations. In this brainstorming activity, you say a word, and students respond by writing the first word that comes to mind. Then students share their words with the class. This activity builds connections to students' prior frames of reference.

Using Science Vocabulary in Context

In order for a new vocabulary word to become part of a student's functional vocabulary, he or she must have ample opportunities to hear and use it. The use of vocabulary terms is embedded in the activities through teacher talk, whole-class and small-group discussions during investigations, writing in science notebooks, readings, assessments, and games. In addition, other methods used during language-arts instruction can be used to reinforce important vocabulary words and phrases.

Word wall/word cards. Use chart paper or a pocket chart to record both science content and procedural words. Record the words as they come up during and after the investigations. Then copy key words on sentence strips or cards, and put them in a pocket chart. With a pocket chart, words can be sorted and moved around easily. For example, you could ask students to find words that are synonyms, antonyms, nouns, or verbs. Word cards should be available to each group during the investigation. This allows students to retrieve a word quickly when they are labeling diagrams and objects used during the investigation.

Drawings and diagrams. For English learners and visual learners, a diagram can be used to review and explain abstract content. Ahead of time, draw an illustration lightly, almost invisibly, with pencil on chart paper. When it's time for the investigation, trace the illustration with markers as you introduce the words and phrases to students. Students will be amazed by your artistic ability.

Cloze activities. Structure a sentence for students to complete, leaving out the vocabulary word, and crafting the sentence so that the missing vocabulary word is the last word in the sentence. You can do this chorally with primary students or in writing on the board or chart paper for upper-elementary students. Here's an example from the **Solids and Liquids Module**.

> Teacher: *Liquids that are clear and that you can see through are _____.*
>
> Students: *Transparent.*

Science Vocabulary Strategies

- *Word wall/word cards*
- *Drawings and diagrams*
- *Cloze activities*
- *Word wizard*
- *Word analysis/word parts*
- *Breaking apart words*
- *Possible sentences*
- *Reading*
- *Highlighting the vocabulary*
- *Index*
- *Poems, chants, and songs*
- *Games*

Science-Centered Language Development

Science-Centered Language Development

Word wizard. Tell students that you are going to lead a word activity. You will be thinking of a science vocabulary word from the word wall. The goal is to figure out the word. Provide hints that have to do with parts of a definition, root word, prefix, suffix, and other relevant components. Students work in teams of two to four. Provide one hint, and give teams 1 minute to discuss. One team member writes the word on a piece of paper or on the whiteboard, using dark marking pens. Each team holds up its word for only you to see. After the third clue, reveal the word, and move on to the next word.

1. *This word is part of a plant.*
2. *It is usually not green.*
3. *It brings water and nutrients into the plant.*

It is the **root**.

Word analysis/word parts. Learning clusters of words that share a common origin can help students understand content-area texts and connect new words to familiar ones. This type of contextualized teaching meets the immediate need of understanding an unknown word while building generative knowledge that supports students in figuring out difficult words for future reading.

geology

geologist

geological

geography

geometry

geophysical

Breaking apart words. Have teams of two to four students break the word into prefix, root word, and suffix. Give each team different words, and have each team share the parsed elements of the word with the whole class.

electromagnetism

electro: having to do with electricity

magnet: having polar properties; attraction and repulsion of opposite and similar poles

ism: relating to a theory of how things behave

Possible sentences. Here is a simple strategy for teaching word meanings and generating class discussion.

1. Choose six to eight key concept words from the text of an article in *FOSS Science Resources*.

2. Choose four to six additional words that students are more likely to know something about.

3. Put the list of 10–14 words on the board or project it. Provide brief definitions as needed.

4. Ask students to devise sentences that include two or more words from the list.

5. On chart paper, write all sentences that students generate, both coherent and otherwise.

6. Have students read the article from which the words were extracted.

7. Revisit students' sentences, and discuss whether the sentences are sensible based on the passage or how they could be modified to be more coherent.

Reading. After the active investigation, students continue to develop their understanding of the vocabulary words and the concepts those words represent by listening to you read aloud, reading with a partner, or reading independently. Use strategies discussed in the Reading Domain section to encourage students to articulate their thoughts and practice the new vocabulary.

Highlighting the vocabulary. Emphasize the vocabulary words students should be using when they answer the focus question in their science notebooks. Distribute copies of the vocabulary/glossary for the investigation (available on FOSSweb) for students to glue into their notebooks. As you introduce the words in the investigation, students highlight them.

Index. Have students create an index at the back of their notebooks. There, they can record new vocabulary words and the notebook page where they defined and used the new words for the first time in the context of the investigation.

Poems, chants, and songs. Vocabulary words and phrases can be reinforced using content-rich poems, rhymes, chants, and songs.

Games. The informal activities included in the investigations are designed to reinforce important vocabulary words. Once students learn them, the words can be integrated into any type of independent work time, such as centers, workshops, and early-finisher tasks.

▶ **NOTE**
See the Science Notebooks in Grades 3–5 chapter for an example of this index.

Science-Centered Language Development

Science-Centered Language Development

> **NOTE**
> English-language development refers to the advancement of students' ability to read, write, and speak English.

> **EL NOTE**
> *Look for EL Notes in the investigations at points where students at beginning levels of English proficiency may need additional support.*

ENGLISH-LANGUAGE DEVELOPMENT

Active investigations, together with ample opportunities to develop and use language, provide an optimal learning environment for English learners. This section highlights the English-language development (ELD) opportunities inherent in FOSS investigations and suggests other best practices for facilitating both the learning of new science concepts and the development of academic vocabulary and language structures that enhance literacy. For example, the hands-on structure of FOSS investigations is essential for the conceptual development of science content knowledge and the habits of mind that guide and define scientific practices. Students are engaged in concrete experiences that are meaningful and that provide a shared context for developing understanding—critical components for ELD instruction.

To further address the needs of English learners, the *Investigations Guide* includes EL Notes at points in the investigations where students at beginning levels of English proficiency may need additional support. When getting ready for an investigation, review the EL Notes, and determine the points where English learners may require scaffolds and where the whole class might benefit from additional language-development supports. One way to plan for ELD integration in science is to keep in mind four key areas: prior knowledge, comprehensible input, academic language development, and oral practice. The ELD chart below lists examples of universal strategies for each of these components that work particularly well in teaching science.

English-Language Development (ELD) Quadrants	
Activating prior knowledge • Inquiry chart • Circle map • Observation poster • Quick write • Kit inventory	**Using comprehensible input** • Content objectives • Multiple exposures • Visual aids • Supported reading • Procedural vocabulary
Developing academic language • Language objectives • Sentence frames • Word wall, word cards, drawings • Concept maps • Cognitive content dictionaries	**Providing oral practice** • Small-group discussions • Science talk • Oral presentations • Poems, chants, and songs • Teacher feedback

Full Option Science System

Activating Prior Knowledge

When an investigation engages a new concept, first students recall and discuss familiar situations, objects, or experiences that relate to and establish a foundation for building new knowledge and conceptual understanding. Eliciting prior knowledge also supports learning by motivating interest, acknowledging culture and values, and checking for misconceptions and prerequisite knowledge. This is usually done in the first steps of Guiding the Investigation in the form of an oral discussion, presentation of new materials, or a written response to a prompt. The tools outlined below can also be used before beginning an investigation to establish a familiar context for launching into new material.

Circle maps. Draw two concentric circles on chart paper. In the middle circle, write the topic to be explored. In the second circle, record what students already know about the subject. Ask students to think about how they know or learned what they already know about the topic. Record the responses outside the circles. Students can also do this independently in their science notebooks.

Strategies for Activating Prior Knowledge
- *Circle maps*
- *Observation posters*
- *Quick writes*
- *Kit inventories*

An example of a circle map

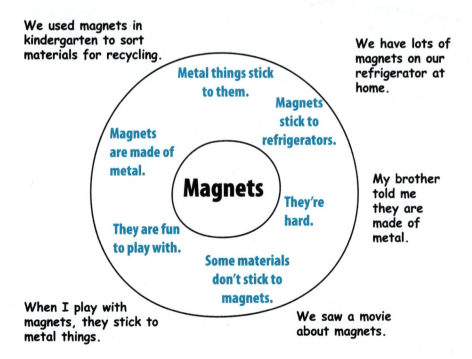

Science-Centered Language Development

Science-Centered Language Development

Observation posters. Make observation posters by gluing or taping pictures and artifacts relevant to the module or a particular investigation onto pieces of blank chart paper or poster paper. Hang them on the walls in the classroom, and have students rotate in small groups to each poster. At each station, students discuss their observations with their partners or small groups and then record (write, draw, or dictate) an observation, a question, a prediction, or an inference about the pictures as a contribution to the commentary on the poster.

TEACHING NOTE

The images selected for the observation posters should pique students' interest but not take away from discoveries that students will make when working with the materials.

An observation poster

Quick writes. Ask students what they think they know about the topic of the investigation. Responses can be recorded independently as a quick write in science notebooks and then shared collaboratively. You should not correct misconceptions initially. Periodically revisit the quick-write ideas as a whole class, or have students review their notebook entries to correct, confirm, or complete their original thoughts as new information is acquired (possibly using a line of learning). At the conclusion of the investigation, students should be able to express mastery of the new conceptual material.

Kit inventories. Introduce each item from the FOSS kit used in the investigation, and ask students questions to get them thinking about what each item is and where they may have seen it before. Have them describe the objects and make predictions about how they will be used in the investigation. Tape samples of the items on chart paper, or print and display the equipment photo cards (download from FOSSweb) along with the name and a description, to serve as an interactive word wall.

Wood cylinder

Item	Teacher	Student
Bottle	What is this? What is it made of? What is it used for?	A bottle. Plastic. To hold liquids.
Scoop	What is this? It's like a spoon. It's called a scoop. What do you think we will be using it for in science?	It looks like a spoon. Like an ice-cream scoop! To scoop things up.
Funnel	Have you seen this before? It's called a funnel. Where else have you seen this? Can you describe it?	My mom uses that for the car. In the kitchen. My uncle uses it sometimes to pour things. It's round on the ends. It's bigger on one end, and it's hollow.
Beaker	Have you seen this before? It's called a beaker. Where else have you seen this?	We used it for science last year to measure and pour water.
Vial	This is a vial. What do you think we will be using it for in science?	To hold small things.

A kit inventory script from the Solids and Liquids Module

Comprehensible Input

In order to initiate their own sense-making process, students must be able to access the information presented to them. We refer to this ability as comprehensible input. Students must understand the essence of new ideas and concepts before beginning the process of constructing new scientific meaning. The strategies for comprehensible input used in FOSS ensure that the delivery of instruction is understandable while providing students with the opportunity to grapple with new ideas and the critically important relationships between concepts. Additional tools such as repetition, visual aids, emphasis on procedural vocabulary, and auditory reinforcement can also be used to enhance comprehensible input for English learners.

Content objectives. The focus question for each investigation part frames the activity objectives—what students should know or be able to do at the end of the part. Making the learning objectives clear and explicit helps English learners prepare to process the delivery of new information, and helps you maintain the focus of the investigation. Write the focus question on the board, have students read it aloud and transcribe it into their science notebooks, and have students answer the focus question at the end of the investigation part. You then check their responses for understanding.

Strategies for Comprehensible Input
- *Content objectives*
- *Multiple exposures*
- *Visual aids*
- *Supported reading*
- *Procedural vocabulary*

Science-Centered Language Development

Procedural Vocabulary
Add
Analyze
Assemble
Attach
Calculate
Change
Classify
Collect
Communicate
Compare
Connect
Construct
Contrast
Describe
Demonstrate
Determine
Draw
Evaluate
Examine
Explain
Explore
Fill
Graph
Identify
Illustrate
Immerse
Investigate
Label
List
Measure
Mix
Observe
Open
Order
Organize
Pour
Predict
Prepare
Record
Represent
Scratch
Separate
Sort
Stir
Subtract
Summarize
Test
Weigh

Multiple exposures. Repeat the activity as a class, at a center during independent work time, or in an analogous but slightly different context, ideally one that incorporates elements that are culturally relevant to students.

Visual aids. On the board or chart paper, write out the steps for conducting the investigation. This will provide a visual reference. Include illustrations if necessary. Use graphic representations (illustrations drawn and labeled in front of students) to review the concepts explored in the active investigations. In addition to the concrete objects included in the kit, use realia to augment the activity to help English learners build understanding and make cultural connections. Graphic organizers (webs, Venn diagrams, T-tables, flowcharts, etc.) aid comprehension by helping students see how ideas (concepts) are related.

Supported reading. In addition to the reading comprehension strategies suggested in the Reading Domain section of this chapter, English learners can also benefit from methods such as front-loading key words, phrases, and complex text structures before reading; using preview-review (main ideas are previewed in the primary language, read in English, and reviewed in the primary language); and having students use sentence frames specifically tailored to record key information and/or graphic organizers that make the content and the relationship between concepts visually explicit from the text as they read.

Procedural vocabulary. Make sure students understand the meaning of the words used in the directions describing what they should be doing during the investigation. These may or may not be science-specific words. Use techniques such as modeling, demonstrating, and body language (gestures) to explain procedural meaning in the context of the investigation. The words students will encounter in FOSS include those listed in the sidebar. To build academic literacy, English learners need to learn the multiple meanings of these words and their specific meanings in the context of science.

Developing Academic Language

As students learn the nuances of the English language, it is critical that they build proficiency in academic language in order to participate fully in the cognitive demands of school. Academic language refers to the more abstract, complex, and specific aspects of language, such as the words, grammatical structure, and discourse markers that are needed for higher cognitive learning. FOSS investigations introduce and provide opportunities for students to practice using the academic vocabulary needed to access and meaningfully engage with science ideas.

Language objectives. Consider the language needs of English learners and incorporate specific language-development objectives that will support learning the science content of the investigation, such as a specific word knowledge skill (a way to expand use of vocabulary by looking at root words, prefixes, and suffixes), a linguistic pattern or structure for oral discussion and writing, or a reading-comprehension strategy. Recording in students' science notebooks is a productive place to optimize science learning and language objectives. For example, in the **Pebbles, Sand, and Silt Module**, one language objective might be "Students will describe what happened when they rubbed their rocks together (think-pair-share) and answer the focus question in their notebooks, using a cause–and–effect sentence frame."

Vocabulary development. The Science-Vocabulary Development section in this chapter describes the ways science vocabulary is introduced and developed in the context of an active investigation and suggests methods and strategies that can be used to support vocabulary development during instruction in English language arts and ELD. In addition to science vocabulary, students also need to learn the nonspecific-content words that facilitate deeper understanding and communication skills. Words such as *release*, *convert*, *beneficial*, *produce*, *receive*, *source*, and *reflect* are words used in the investigations and *FOSS Science Resources* and are frequently used in other content areas. Learning these academic vocabulary words gives students a more precise and complex way of practicing and communicating productive thinking. Consider using the strategies described in the Science-Vocabulary Development section to explicitly teach targeted, high-leverage words that can be used in multiple ways and that can help students make connections to other words and concepts. Sentence frames, word wall, concept maps, and cognitive content dictionary are strategies that have been found to be effective with academic-vocabulary development.

Science-Centered Language Development

Science-Centered Language Development

> **NOTE**
> For additional resources and updated references, go to FOSSweb.

REFERENCES

Applebee, A. 1984. "Writing and Reasoning." *Review of Educational Research* 54 (winter): 577–596.

Bereiter, C. 2002. *Education and Mind in the Knowledge Age*. Hillsdale, NJ: Erlbaum.

Black Hills Special Services Cooperative. 2006. "3-2-1 Strategy." In *On Target: More Strategies to Guide Learning*. Rapid City, SD. http://www.sdesa6.org/content/resources.htm.

Dyasi, H. M. 2006. "Visions of Inquiry: Science." In *Linking Science and Literacy in the K–8 Classroom*, ed. R. Douglas, K. Worth, and W. Binder. Arlington, VA: NSTA Press.

Gibbons, P. 2002. *Scaffolding Language, Scaffolding Learning*. Portsmouth, NH: Heinemann.

Graham, S., and M. Herbert. 2010. *Writing to Read: Evidence for How Writing Can Improve Reading*. New York: Carnegie.

Hamilton, G. 2002. *Content-Area Reading Strategies: Science*. Portland, ME: Walch Publishing.

Hand, B., and V. Prain. 2002. "Teachers Implementing Writing-to-Learn Strategies in Junior Secondary Science: A Case Study." *Science Education* 86: 737–755.

Harvey, S. 1998. *Nonfiction Matters: Reading, Writing, and Research in Grades 3–8*. Portland, ME: Stenhouse.

Harvey, S., and A. Goudvis. 2007. *Strategies That Work: Teaching Comprehension for Understanding and Engagement*. Portland, ME: Stenhouse.

Keene, E., and S. Zimmermann. 2007. *Mosaic of Thought: The Power of Comprehension Strategies*. 2nd ed. Portsmouth, NH: Heinemann.

Keys, C. 1999. *Revitalizing Instruction in Scientific Genres: Connecting Knowledge Production with Writing to Learn in Science*. Athens: University of Georgia.

Lieberman, G. A., and L. L. Hoody. 1998. *Closing the Achievement Gap: Using the Environment as an Integrating Context for Learning*. San Diego, CA: State Education and Environment Roundtable.

National Research Council. 2012. *A Framework for K–12 Science Education: Practices, Crosscutting Concepts, and Core Ideas*. Committee on Conceptual Framework for New Science Education Standards.

Ostlund, K. 1998. "What the Research Says about Science Process Skills: How Can Teaching Science Process Skills Improve Student Performance in Reading, Language Arts, and Mathematics?" *Electronic Journal of Science Education* 2 (4).

Wellington, J., and J. Osborne. 2001. *Language and Literacy in Science Education.* Buckingham, UK: Open University Press.

Winokur, J., and K. Worth. 2006. "Talk in the Science Classroom: Looking at What Students and Teachers Need to Know and Be Able to Do." In *Linking Science and Literacy in the K–8 Classroom*, ed. R. Douglas, K. Worth, and W. Binder. Arlington, VA: NSTA Press.

Science-Centered Language Development

FOSS and Common Core ELA – Grade 5

FOSS and Common Core ELA — Grade 5

Contents

Introduction 1

Reading Standards for
Informational Text 4

Reading Standards:
Foundational Skills 8

Writing Standards 10

Speaking and Listening
Standards 16

Language Standards 20

INTRODUCTION

Each FOSS investigation follows a similar design to provide multiple exposures to science concepts. The design includes these pedagogies.

- Active investigation, including outdoor experiences
- Writing in science notebooks to answer focus questions
- Reading in *FOSS Science Resources*
- Assessment to monitor progress and motivate student reflection on learning

In practice, these components are seamlessly integrated into a continuum designed to maximize every student's opportunity to learn. An instructional sequence may move from one pedagogy to another and back again to ensure adequate coverage of a concept.

The FOSS instructional design recognizes the important role of language in science learning. Throughout the pedagogical design elements, students engage in the practices of the Common Core State Standards for English Language Arts. The purpose of this chapter is to provide the big picture of how FOSS provides opportunities for the development and exercising of these practices through science. On the following pages, there is a chart that identifies the opportunities for fifth grade and where the relevant opportunities are found within the three FOSS modules.

FOSS and Common Core ELA — Grade 5

Guiding Principles

When integrating language-arts instruction with FOSS, keep in mind these guiding principles:

- FOSS investigations follow a clear and coherent conceptual flow and a consistent instructional design. Students develop science knowledge by building a framework of concepts and supporting ideas.

- Common Core State Standards for ELA are introduced, developed, and practiced in the context of learning science content and engaging in the science and engineering practices. Students read and comprehend complex science texts related to their prior experience and knowledge. They write informational/explanatory texts, arguments to support claims, and narratives about experiences in science. They engage in collaborative discussions about science and learn new vocabulary and language structures in context.

- The decision to use additional science texts, writing tasks, oral discourse opportunities, and vocabulary development activities is based on how well they address the science as well as the ELA standards.

- Instruction is differentiated to meet the needs of all students; the linguistic accommodations that are made for English learners support comprehensible input and accelerate academic language development. Language objectives for English learners in science instruction include the application of strategies that support construction of meaning from academic discussions and complex text, participation in productive discourse, and the ability to express ideas in writing clearly and coherently according to task, purpose, and audience.

- Formative assessment tools are used routinely to measure progress toward science understanding, use of science and engineering practices, and meeting literacy and language development goals. Assessment is viewed as a way to make student thinking visible and to determine next steps for instruction for both science and literacy. Instruction includes opportunities for students to assess themselves and peers.

Adhering to these guiding principles optimizes instructional time and, most importantly, benefits student learning by providing authentic and relevant contexts for building content knowledge, applying meaning-making strategies, and developing language and literacy skills.

Fifth grade is a critical year for students as they consolidate their literacy skills and apply them across content areas and in different settings. Fifth graders read widely and deeply from a range of challenging informational texts that support their content learning and expand their vocabulary. They communicate in complex and flexible ways that demonstrate understanding of task, purpose and audience. Fifth graders are expected to use the science and engineering practices to demonstrate their understanding of the core ideas. To accomplish this, students apply language structures for sequencing, comparing and contrasting, determining cause-and-effect relationships, and problem-solving.

Instructional Flow

In almost all investigations, the instructional flow is the same and provides these opportunities for effective integration of ELA standards.

- When **setting the context** for the lesson, students activate prior knowledge through class or small-group discussions where they report on a topic or present an opinion using appropriate facts and relevant, descriptive details (SL 4).

- During the **active investigation**, students are expected to work with partners and in collaborative groups, and to engage in teacher-led discussions where they build on each other's ideas and express their own clearly (SL 1).

- In the **data management** phase, students make observations, and then routinely record and organize data in their notebooks (W 10). The notebook provides a space for students to recall information from experience, gather information from print and other media, summarize or paraphrase information in notes (W 8) and to acquire and use general academic and domain-specific words and phrases (L 6).

- The **analysis** phase involves discussing data, constructing and writing explanations, and engaging in argumentation. Here, students are making meaning by writing explanatory texts (W 2), writing opinion pieces supporting a point of view with reasons (W 1), or conducting short research projects that build knowledge through investigation of different aspects of a topic. (W 7).

- **Reading** articles in *FOSS Science Resources* and other recommended readings provides a plethora of opportunities to address all the fifth-grade reading standards for informational text.

- Lastly, the **assessment** tools and next-step strategies for engaging students in high-level critical thinking support the development of the Common Core State Standards capacities of the literate individual: demonstrate independence, build strong content knowledge, comprehend as well as critique, and value evidence.

Again, we have provided you with some examples of how FOSS connects to the fifth-grade ELA standards; there are many more opportunities waiting to be created and explored by you and your students.

> **TEACHING NOTE**
>
> *Throughout the fifth-grade FOSS modules, opportunities for addressing the ELA standards have been noted; however these examples should not be considered the only places for integrating literacy skills.*

FOSS and Common Core ELA — Grade 5

READING STANDARDS FOR INFORMATIONAL TEXT

	Grade 5 Standard	Mixtures and Solutions Module
Key Ideas and Details	1. Quote accurately from a text when explaining what the text says explicitly and when drawing inferences from the text.	Discuss articles in *FOSS Science Resources* Inv 1, Part 2, Steps 22, 23 Inv 2, Part 2, Step 12; Inv 2, Part 3, Steps 10, 17 Inv 3, Part 2, Step 20; Inv 3, Part 3, Step 10 Inv 3, Part 4, Steps 15-17 Inv 4, Part 3, Steps 22, 23; Inv 4, Part 4, Steps 15, 26 Inv 5, Part 2, Step 17
	2. Determine two or more main ideas of a text and explain how they are supported by key details; summarize the text.	Discuss and review articles in *FOSS Science Resources* Inv 1, Part 2, Step 22; Inv 1, Part 3, Step 15 Inv 2, Part 2, Step 12 Inv 3, Part 1, Step 18; Inv 3, Part 2, Step 20 Inv 4, Part 1, Steps 23, 24; Inv 4, Part 3, Steps 22, 23 Inv 5, Part 1, Step 21; Inv 5, Part 2, Steps 17, 18
	3. Explain the relationships or interactions between two or more individuals, events, ideas, or concepts in a historical, scientific, or technical text based on specific information in the text.	Discuss articles in *FOSS Science Resources* Inv 1, Part 2, Step 22 Inv 2, Part 2, Step 11; Inv 2, Part 3, Step 10 Inv 3, Part 1, Step 17; Inv 3, Part 2, Step 20 Inv 4, Part 1, Steps 23, 24; Inv 4, Part 3, Steps 22, 23 Inv 5, Part 2, Steps 17, 18
Craft and Structure	4. Determine the meaning of general academic and domain-specific words and phrases in a text relevant to a *grade 5 topic or subject area*.	All investigations provide opportunities for students to determine the meaning of academic and science-specific words and phrases while reading. Inv 1, Part 2, Step 22 Inv 2, Part 2, Step 12 Inv 3, Part 1, Step 18; Inv 3, Part 2, Step 20
	5. Compare and contrast the overall structure (e.g., chronology, comparison, cause/effect, problem/solution) of events, ideas, concepts, or information in two or more texts.	Compare and contrast text structure in *FOSS Science Resources* with other texts Inv 3, Part 2, Step 19 Inv 4, Part 1, Step 22 Inv 4, Part 3, Step 21 Inv 5, Part 2, Step 18
	6. Analyze multiple accounts of the same event or topic, noting important similarities and differences in the point of view they represent.	Read and compare article topics in *FOSS Science Resources* to other texts. Selected examples Inv 2, Part 2, Step 12; Inv 3, Part 3, Step 10; Inv 4, Part 3, Step 22; Inv 4, Part 4, Steps 25, 27

Common Core State Standards for English language arts and literacy in history/social studies science and technical subjects (National Governors Association Center for Best Practices and Council of Chief State School Officers, 2010).

Earth and Sun Module	Living Systems Module
Discuss articles in *FOSS Science Resources* Inv 1, Part 2, Steps 20, 21; Inv 1, Part 3, Steps 24, 25 Inv 2, Part 1, Steps 16, 17, 21; Inv 2, Part 2, Steps 11, 14 Inv 2, Part 4, Steps 21, 28; Inv 2, Part 5, Step 18 Inv 3, Part 1, Step 19; Inv 3, Part 2, Step 11 Inv 4, Part 2, Step 25; Inv 4, Part 3, Step 27 Inv 5, Part 3, Steps 23, 26; Inv 5, Part 4, Steps 13, 18	Discuss articles in *FOSS Science Resources* Inv 1, Part 1, Step 11; Inv 1, Part 2, Steps 6, 25, 26 Inv 1, Part 3, Steps 17, 19; Inv 1, Part 4, Step 16 Inv 2, Part 1, Step 24; Inv 2, Part 2, Step 8 Inv 3, Part 1, Steps 12, 33, 36; Inv 3, Part 2, Steps 8-9 Inv 4, Part 1, Step 7; Inv 4, Part 2, Steps 15, 17 Inv 4, Part 3, Step 11; Inv 4, Part 4, Step 11
Discuss and review articles in *FOSS Science Resources* Inv 1, Part 2, Steps 20, 21; Inv 1, Part 3, Steps 24, 25 Inv 2, Part 1, Steps 15, 20; Inv 2, Part 2, Step 14 Inv 3, Part 1, Steps 18, 19; Inv 3, Part 2, Step 8 Inv 4, Part 1, Step 25; Inv 4, Part 2, Step 24 Inv 5, Part 1, Steps 21, 25; Inv 5, Part 4, Steps 12, 13, 17	Discuss and review articles in *FOSS Science Resources* Inv 1, Part 1, Step 10; Inv 1, Part 2, Steps 5, 6, 25, 26 Inv 1, Part 3, Step 16; Inv 1, Part 4, Step 15 Inv 2, Part 1, Step 23; Inv 2, Part 2, Step 9 Inv 3, Part 1, Steps 12, 33, 36; Inv 3, Part 2, Steps 7, 10 Inv 4, Part 1, Steps 5, 6; Inv 4, Part 2, Steps 14-17
Discuss articles in *FOSS Science Resources* Inv 1, Part 2, Steps 20, 21; Inv 1, Part 3, Steps 24, 25 Inv 2, Part 1, Steps 16, 21; Inv 2, Part 2, Steps 11, 14 Inv 2, Part 4, Steps 21, 28; Inv 2, Part 5, Steps 17, 18, 23 Inv 3, Part 2, Step 11 Inv 4, Part 1, Steps 24, 25; Inv 4, Part 2, Steps 24, 25 Inv 5, Part 1, Steps 22, 23; Inv 5, Part 3, Steps 23, 26	Discuss articles in *FOSS Science Resources* Inv 1, Part 2, Steps 5, 25, 26; Inv 1, Part 3, Steps 16, 17 Inv 2, Part 1, Step 24; Inv 2, Part 2, Step 9 Inv 3, Part 1, Steps 12, 32, 33, 36; Inv 3, Part 2, Step 10; Inv 3, Part 3, Steps 17, 18 Inv 4, Part 1, Step 7; Inv 4, Part 2, Steps 15, 17; Inv 4, Part 3, Step 11; Inv 4, Part 4, Step 11
All investigations provide opportunities for students to determine the meaning of academic and science-specific words and phrases while reading. Inv 2, Part 2, Step 14; Inv 2, Part 5, Step 23 Inv 3, Part 1, Steps 18, 19; Inv 3, Part 3, Step 18 Inv 4, Part 2, Step 25; Inv 4, Part 3, Step 27 Inv 5, Part 3, Step 23, 26; Inv 5, Part 4, Step 12	All investigations provide opportunities for students to determine the meaning of academic and science-specific words and phrases while reading. Inv 1, Part 1, Step 11; Inv 1, Part 2, Step 27 Inv 2, Part 1, Step 23; Inv 2, Part 2, Step 9 Inv 3, Part 1, Steps 12, 31, 32, 33; Inv 3, Part 2, Step 7 Inv 4, Part 1, Step 5; Inv 4, Part 2, Step 18
Compare and contrast text structure in *FOSS Science Resources* with other texts Inv 2, Part 1, Step 21; Inv 2, Part 4, Step 21 Inv 3, Part 3, Step 18 Inv 4, Part 3, Step 28; Inv 4, Part 4, Step 27 Inv 5, Part 4, Step 13	Compare and contrast text structure in *FOSS Science Resources* with other texts Inv 1, Part 3, Step 17 Inv 3, Part 1, Step 36 Inv 4, Part 3, Step 11
Read and compare article topics in *FOSS Science Resources* to other texts. Selected examples Inv 2, Part 1, Step 21; Inv 2, Part 2, Step 11 Inv 2, Part 4, Step 21; Inv 2, Read Moon myths	Read and compare article topics in *FOSS Science Resources* to other texts. Selected examples Inv 1, Part 3, Step 17 Inv 4, Part 3, Step 11

FOSS and Common Core ELA — Grade 5

READING STANDARDS FOR INFORMATIONAL TEXT

	Grade 5 Standard	Mixtures and Solutions Module
Integration of Knowledge and Ideas	7. Draw on information from multiple print or digital sources, demonstrating the ability to locate an answer to a question quickly or to solve a problem efficiently.	Use *FOSS Science Resources* to locate information. Selected examples Inv 1, Part 2, Step 22 Inv 2, Part 2, Step 12; Inv 2, Part 3, Step 10 Inv 4, Part 3, Step 22; Inv 4, Part 4, Step 24 Inv 5, Part 2, Steps 7, 17
	8. Explain how an author uses reasons and evidence to support particular points in a text, identifying which reasons and evidence support which point(s).	Read and discuss articles in *FOSS Science Resources* Inv 1, Part 2, Step 22 Inv 3, Part 1, Step 18; Inv 3, Part 3, Step 9 Inv 4, Part 1, Step 24; Inv 4, Part 4, Step 25
	9. Integrate information from several texts on the same topic in order to write or speak about the subject knowledgeably.	Students can read their *FOSS Science Resources* as well as readings suggested on FOSSweb. Using several texts, students integrate information when speaking and writing about science content. Selected examples Inv 1, Part 4, Step 16 Inv 1, Language Extensions Inv 3, Part 3, Step 13; Inv 3, Part 4, Step 17 Inv 4, Part 3, Step 22; Inv 4, Part 4, Steps 25, 27
Range of Reading and Level of Text Complexity	10. By the end of the year, read and comprehend informational texts, including history/social studies, science, and technical texts, at the high end of the grades 4–5 text complexity band independently and proficiently.	All investigations provide opportunities for students to develop their ability to read and comprehend complex informational science text such as *FOSS Science Resources*.

Earth and Sun Module	Living Systems Module
Use *FOSS Science Resources* to locate information. Selected examples Inv 2, Part 2, Step 13; Inv 2, Part 5, Steps 18, 20, 21 Inv 3, Part 2, Step 11 Inv 4, Part 2, Steps 23, 24; Inv 4, Part 3, Step 26 Inv 5, Part 1, Step 22; Inv 5, Part 3, Step 23 Inv 5, Part 4, Step 12	Use *FOSS Science Resources* to locate information. Selected examples Inv 1, Part 2, Steps 5, 25, 26; Inv 1, Part 3, Steps 16, 17 Inv 1, Part 4, Steps 14, 15 Inv 2, Part 1, Step 23; Inv 2, Part 2, Step 9 Inv 2, Part 3, Steps 7, 10 Inv 3, Part 1, Steps 12, 32, 33, 36; Inv 3, Part 2, Step 10; Inv 3, Part 3, Steps 17, 18 Inv 4, Part 2, Steps 14-18; Inv 4, Part 3, Step 11
Read and discuss articles in *FOSS Science Resources* Inv 1, Part 3, Steps 24, 25 Inv 2, Part 2, Step 14 Inv 3, Part 2, Step 11 Inv 5, Part 4, Step 17	Read and discuss articles in *FOSS Science Resources* Inv 1, Part 2, Step 4; Inv 1, Part 3, Step 19 Inv 2, Part 2, Step 9; Inv 2, Part 3, Step 9 Inv 3, Part 1, Step 33 Inv 4, Part 1, Step 6
Students can read their *FOSS Science Resources* as well as readings suggested on FOSSweb. Using several texts, students integrate information when speaking and writing about science content. Selected examples Inv 2, Part 2, Step 11 Inv 3, Part 3, Step 18 Inv 4, Part 3, Step 28; Inv 4, Part 4, Step 27 Inv 5, Part 3, Step 26; Inv 5, Part 4, Steps 13, 18	Students can read their *FOSS Science Resources* as well as readings suggested on FOSSweb. Using several texts, students integrate information when speaking and writing about science content. Selected examples Inv 1, Part 2, Step 26; Inv 1, Part 3, Step 17 Inv 2, Science Extensions. Read about the man with the hole in his stomach Inv 3, Part 1, Step 36
All investigations provide opportunities for students to develop their ability to read and comprehend complex informational science text such as *FOSS Science Resources*.	All investigations provide opportunities for students to develop their ability to read and comprehend complex informational science text such as *FOSS Science Resources*.

FOSS and Common Core ELA — Grade 5

READING STANDARDS: FOUNDATIONAL SKILLS

	Grade 5 Standard	Mixtures and Solutions Module
Phonics and Word Recognition	3. Know and apply grade-level phonics and word analysis skills in decoding words. a. Use combined knowledge of all letter-sound correspondences, syllabication patterns, and morphology (e.g., roots and affixes) to read accurately unfamiliar multisyllabic words in context and out of context.	All investigations provide opportunities for students to apply decoding skills while reading articles in *FOSS Science Resources*. Selected example Inv 2, Part 2, Step 10
Fluency	4. Read with sufficient accuracy and fluency to support comprehension. a. Read grade-level text with purpose and understanding. b. Read grade-level prose and poetry orally with accuracy, appropriate rate, and expression on successive readings. c. Use context to confirm or self-correct word recognition and understanding, rereading as necessary.	All investigations provide opportunities for students to practice reading with accuracy and fluency. Selected examples Inv 1, Part 2, Step 22; Inv 1, Part 3, Step 15 Inv 1, Part 4, Step 14 Inv 2, Part 2, Step 11; Inv 2, Part 3, Steps 9, 16 Inv 3, Part 1, Step 18; Inv 3, Part 2, Step 19 Inv 3, Part 3, Steps 9, 12; Inv 3, Part 4, Step 14 Inv 4, Part 1, Step 23; Inv 4, Part 3, Step 21 Inv 4, Part 4, Steps 14, 23, 26 Inv 5, Part 1, Step 20; Inv 5, Part 2, Step 16

Earth and Sun Module	Living Systems Module
All investigations provide opportunities for students to apply decoding skills while reading articles in *FOSS Science Resources*. Selected example Inv 2, Part 2, Step 14	All investigations provide opportunities for students to apply decoding skills while reading articles in *FOSS Science Resources*. Selected example Inv 2, Part 2, Step 9
All investigations provide opportunities for students to practice reading with accuracy and fluency. Selected examples Inv 1, Part 2, Step 21; Inv 1, Part 3, Step 24 Inv 2, Part 2, Step 12; Inv 2, Part 4, Steps 20, 27 Inv 2, Part 5, Steps 17, 21, 23 Inv 3, Part 1, Step 18; Inv 3, Part 2, Step 10 Inv 3, Part 3, Step 17 Inv 4, Part 1, Step 23; Inv 4, Part 2, Step 23 Inv 4, Part 3, Step 26; Inv 4, Part 4, Step 26 Inv 5, Part 1, Step 21; Inv 5, Part 3, Steps 21, 25 Inv 5, Part 4, Steps 12, 17	All investigations provide opportunities for students to practice reading with accuracy and fluency. Selected examples Inv 1, Part 1, Step 10; Inv 1, Part 2, Steps 5, 25 Inv 1, Part 3, Step 16; Inv 1, Part 4, Step 14 Inv 2, Part 1, Step 22; Inv 2, Part 2, Step 8 Inv 2, Part 3, Steps 5, 9 Inv 3, Part 1, Steps 12, 32-35; Inv 3, Part 2, Steps 6-9 Inv 4, Part 2, Steps 14, 16

FOSS and Common Core ELA — Grade 5

WRITING STANDARDS

Grade 5 Standard	Mixtures and Solutions Module
1. Write opinion pieces on topics or texts, supporting a point of view with reasons and information. a. Introduce a topic or text clearly, state an opinion, and create an organizational structure in which ideas are logically grouped to support the writer's purpose. b. Provide logically ordered reasons that are supported by facts and details. c. Link opinion and reasons using words, phrases, and clauses (e.g., *consequently, specifically*). d. Provide a concluding statement or section related to the opinion presented.	All investigations provide opportunities for students to write their opinion, or claim, supported by reasons. Students answer questions (focus questions, response sheets, assessments) by stating their claim supported by evidence and reasoning. Selected examples Inv 1, Part 2, Step 9; Inv 1, Part 4, Step 11 Inv 4, Part 1, Step 20; Inv 4, Part 2, Step 12 Inv 4, Part 3, Steps 11, 17 Inv 5, Part 1, Step 18; Inv 5, Part 2, Step 15
2. Write informative/explanatory texts to examine a topic and convey ideas and information clearly. a. Introduce a topic clearly, provide a general observation and focus, and group related information logically; include formatting (e.g., headings), illustrations, and multimedia when useful to aiding comprehension. b. Develop the topic with facts, definitions, concrete details, quotations, or other information and examples related to the topic. c. Link ideas within and across categories of information using words, phrases, and clauses (e.g., *in contrast, especially*). d. Use precise language and domain-specific vocabulary to inform about or explain the topic. e. Provide a concluding statement or section related to the information or explanation presented.	All investigations provide opportunities for students to write explanatory texts to examine the science topic they are learning. In every part, students write an explanation as part of their answer to the focus question or the response sheet. Selected examples Inv 1, Part 4, Step 18 Inv 3, Part 1, Step 14; Inv 3, Part 2, Step 17 Inv 3, Part 4, Steps 10, 18 Inv 4, Part 4, Steps 20, 29
3. Write narratives to develop real or imagined experiences or events using effective technique, descriptive details, and clear event sequences. a. Orient the reader by establishing a situation and introducing a narrator and/or characters; organize an event sequence that unfolds naturally. b. Use narrative techniques, such as dialogue, description, and pacing, to develop experiences and events or show the responses of characters to situations. c. Use a variety of transitional words, phrases, and clauses to manage the sequence of events. d. Use concrete words and phrases and sensory details to convey experiences and events precisely. e. Provide a conclusion that follows from the narrated experiences or events.	All investigations provide opportunities for students to write narratives. Students describe their observations and experiences with the science ideas they are exploring. Selected examples Inv 1, Part 3, Step 13 Inv 2, Language Extensions. Write procedures for construction Inv 5, Language Extensions. Describe the reaction. Apply the reaction

Text Types and Purposes

Earth and Sun Module	**Living Systems Module**
All investigations provide opportunities for students to write their opinion, or claim, supported by reasons. Students answer questions (focus question, response sheets, assessments) by stating their claim supported by evidence and reasoning. Selected example Inv 4, Part 4, Step 24	All investigations provide opportunities for students to write their opinion, or claim, supported by reasons. Students answer questions (focus question, response sheets, assessments) by stating their claim supported by evidence and reasoning. Selected examples Inv 1, Part 2, Step 28 Inv 2, Part 2, Step 8, 9; Inv 2, Part 3, Step 17
All investigations provide opportunities for students to write explanatory texts to examine the science topic they are learning. In every part, students write an explanation as part of their answer to the focus question or the response sheet. Selected examples Inv 1, Part 1, Step 15; Inv 1, Part 2, Step 22 Inv 1, Part 3, Steps 25, 29 Inv 2, Part 2, Step 9 Inv 3, Part 3, Step 15; Inv 3, Part 3, Step 19 Inv 4, Part 1, Step 21; Inv 4, Part 2, Step 27 Inv 4, Part 3, Step 23; Inv 4, Part 4, Step 28 Inv 5, Part 1, Step 18; Inv 5, Part 2, Steps 8, 9	All investigations provide opportunities for students to write explanatory texts to examine the science topic they are learning. In every part, students write an explanation as part of their answer to the focus question or the response sheet. Selected examples Inv 1, Part 1, Step 12; Inv 1, Part 3, Steps, 12, 14; Inv 1, Part 4, Step 18 Inv 2, Part 1, Step 26; Inv 2, Part 2, Step 14 Inv 2, Part 3, Steps 16, 24 Inv 3, Part 1, Step 39; Inv 3, Part 2, Step 16 Inv 3, Part 3, Step 14; Inv 3, Language Extension. Write about making maple syrup Inv 4, Part 2, Steps 19, 20; Inv 4, Part 3, Step 7 Inv 4, Part 4, Step 12
All investigations provide opportunities for students to write narratives. Students describe their observations and experiences with the science ideas they are exploring. Selected examples Inv 1, Language Extensions. Go on a treasure hunt, Describe shadows	All investigations provide opportunities for students to write narratives. Students describe their observations and experiences with the science ideas they are exploring. Selected examples Inv 4, Language Extensions. Write captions for pictures

FOSS and Common Core ELA — Grade 5

FOSS and Common Core ELA — Grade 5

WRITING STANDARDS (CONT.)

	Grade 5 Standard	Mixtures and Solutions Module
Production and Distribution of Writing	4. Produce clear and coherent writing in which the development and organization are appropriate to task, purpose, and audience. (Grade-specific expectations for writing types are defined in standards 1–3 above.)	All investigations provide opportunities for students to record and organize their data in their science notebooks. Based on their data, students construct and write their explanations. Selected examples Inv 1, Part 1, Step 19; Part 2, Step 13 Inv 1, Part 3, Steps 6, 13; Inv 1, Part 4, Steps 11, 15, 18 Inv 1, Language Extension. Invent a gorp recipe Inv 2, Part 3, Step 14 Inv 2, Language Extension. Write procedures for construction Inv 3, Part 1, Step 14; Inv 3, Part 3, Step 2 Inv 3, Part 4, Step 10 Inv 4, Part 1, Step 20; Inv 4, Part 2, Steps 12, 16 Inv 4, Part 3, Step 17; Inv 4, Part 4, Step 20 Inv 5, Part 3, Step 14
	5. With guidance and support from peers and adults, develop and strengthen writing as needed by planning, revising, editing, rewriting, or trying a new approach. (Editing for conventions should demonstrate command of Language standards 1–3 up to and including grade 5.)	The Wrap-up/Warm-up section of each investigation part provides the opportunity for students to strengthen their notebook entries by revising and adding in new information. Inv 1, Part 3, Step 23; Inv 1, Part 4, Step 17 Inv 2, Part 3, Steps 13, 17 Inv 3, Part 3, Step 14 Inv 4, Part 3, Step 22 The Wrap-up review focus question section at the end of each investigation and next-step strategies after answering the response sheets or taking the I-Check also serve as a method for strengthening writing.
	6. With some guidance and support from adults, use technology, including the Internet, to produce and publish writing as well as to interact and collaborate with others; demonstrate sufficient command of keyboarding skills to type a minimum of two pages in a single sitting.	

Earth and Sun Module	Living Systems Module
All investigations provide opportunities for students to record and organize their data in their science notebooks. Based on their data, students construct and write their explanations. Selected examples Inv 1, Part 1, Step 15; Inv 1, Part 2, Step 19 Inv 3, Part 3, Step 15 Inv 4, Part 1, Step 21; Inv 4, Part 4, Step 24 Inv 5, Part 2, Step 8	All investigations provide opportunities for students to record and organize their data in their science notebooks. Based on their data, students construct and write their explanations. Selected examples Inv 1, Part 1, Step 12; Inv 1, Part 2, Step 28 Inv 2, Part 1, Step 26; Inv 2, Part 3, Steps 16, 17, 24 Inv 3, Part 2, Step 16; Inv 3, Part 3, Step 14
The Wrap-up/Warm-up section of each investigation part provides the opportunity for students to strengthen their notebook entries by revising and adding in new information. Inv 1, Part 2, Step 25 Inv 3, Part 1, Step 20 Inv 4, Part 1, Step 27; Inv 4, Part 2, Step 29 The Wrap-up review focus question section at the end of each investigation and next-step strategies after answering the response sheets or taking the I-Check also serve as a method for strengthening writing.	The Wrap-up/Warm-up section of each investigation part provides the opportunity for students to strengthen their notebook entries by revising and adding in new information. Inv 1, Part 1, Step 14; Inv 1, Part 2, Step 30; Inv 1, Part 3, Step 21 Inv 2, Part 1, Step 28; Inv 2, Part 2, Step 19 Inv 4, Part 4, Steps 4, 8 The Wrap-up review focus question section at the end of each investigation and next-step strategies after answering the response sheets or taking the I-Check also serve as a method for strengthening writing.
Inv 5, Science and Engineering Extension. Search for severe weather	Inv 2, Part 1, Step 24 Inv 3, Part 1, Step 12

FOSS and Common Core ELA — Grade 5

FOSS and Common Core ELA — Grade 5

WRITING STANDARDS (CONT.)

Grade 5 Standard	Mixtures and Solutions Module
Research to Build and Present Knowledge 7. Conduct short research projects that use several sources to build knowledge through investigation of different aspects of a topic.	All investigations provide opportunities for students to further investigate different aspects of the science topic. Inv 1, Science Extensions. Research diatomaceous earth. Research sodium chloride; Inv 1, Engineering Extension. Engineering without borders Inv 2, Part 2, Step 12; Inv 2, Part 3, Step 10 Inv 3, Part 3, Step 13 Inv 3, Science Extensions. Investigate drinks Inv 4, Part 3, Step 19; Inv 4, Part 4, Steps 12, 25, 27 Inv 4, Language Extensions. Find citric acid; Inv 4, Science Extensions. Change the temperature Inv 5, Science and Engineering Extensions. Investigate baking powder and baking soda
8. Recall relevant information from experiences or gather relevant information from print and digital sources; summarize or paraphrase information in notes and finished work, and provide a list of sources.	All investigations provide students with the opportunity to write and record their observations in their science notebooks. Students also take notes and organize information when reading articles. Selected examples Inv 1, Part 1, Steps 4, 7, 19; Inv 1, Part 2, Steps 6, 7, 21 Inv 1, Part 3, Steps 14, 15; Inv 1, Part 4, Steps 16, 18 Inv 2, Part 1, Steps 5, 28, 11; Inv 2, Part 3, Step 3 Inv 2, Part 3, Steps 6, 9, 16 Inv 3, Part 1, Step 12; Inv 3, Part 2, Steps 16, 18 Inv 3, Part 3, Steps 2, 6, 8, 9; Inv 3, Part 4, Steps 10, 14 Inv 4, Part 3, Steps 4, 6, 9, 11, 21; Inv 4, Part 4, Step 14 Inv 5, Part 1, Step 12; Inv 5, Part 2, Steps 13, 17 Inv 5, Part 3, Steps 3, 5, 8, 18, 19
Range of Writing 9. Draw evidence from literary or informational texts to support analysis, reflection, and research. a. Apply *grade 5 Reading standards* to literature (e.g., "Compare and contrast two or more characters, settings, or events in a story or a drama, drawing on specific details in the text [e.g., how characters interact]"). b. Apply *grade 5 Reading standards* to informational texts (e.g., "Explain how an author uses reasons and evidence to support particular points in a text, identifying which reasons and evidence support which point[s]").	All investigations provide opportunities to use the *FOSS Science Resources* as a source from which to draw evidence to support their ideas (e.g., discussion questions at the end of the articles). Selected examples Inv 1, Part 2, Step 22 Inv 1, Part 4, Step 18 Inv 2, Part 3, Step 17 Inv 4, Part 3, Step 21 Inv 5, Part 3, Step 19

Earth and Sun Module	Living Systems Module
All investigations provide opportunities for students to further investigate different aspects of the science topic. Inv 1, Science and Engineering Extensions. Research sundials; Inv 1, Social Studies Extensions. Research shadow theater Inv 2, Part 1, Steps 16, 21; Inv 2, Part 5, Step 23 Inv 2, Language Extensions. Research Apollo missions. Inv 2, Science Extensions. Investigate eclipses, Research the moons of other planets Inv 3, Part 3, Step 18; Inv 3, Language Extensions. Explore weather topics, Research weather lore Inv 3, Science Extensions. Track weather reports. Find out how digital weather stations work. Inv 5, Science and Engineering Extensions. USDA plant hardiness zone map.	All investigations provide opportunities for students to further investigate different aspects of the science topic. Inv 1, Science Extension. Research vermicomposting Inv 2, Language Extension. Find sugars in products; Inv 2, Science Extensions. Test sugar content of breakfast cereals, Research other digestive organs, Research dialysis Inv 3, Part 1, Step 36; Inv 3, Science Extensions. Investigate flowers, Research asthma, Find out about hearts of other animals
All investigations provide students with the opportunity to write and record their observations in their science notebooks. Students also take notes and organize information when reading articles. Selected examples Inv 1, Part 1, Steps 2, 7, 15 Inv 1, Part 2, Steps 19, 2; Inv 1, Part 3, Steps 23-26 Inv 2, Part 1, Steps 7, 10, 15, 20; Inv 2, Part 4, Steps 11, 18, 27; Inv 2, Part 5, Steps 7, 11, 15, 18, 23 Inv 3, Part 1, Steps 6, 12, 16-18; Inv 3, Part 2, Steps 7, 8, 10; Inv 3, Part 3, Steps 1, 10, 13, 15, 17 Inv 4, Part 1, Steps 6, 15, 21, 24; Inv 4, Part 2, Step 1 Inv 4, Part 3, Steps 10, 14, 20, 25, 28 Inv 4, Part 4, Steps 10, 17 Inv 5, Part 1, Steps 12, 13, 18, 21; Inv 5, Part 2, Steps 8, 9	All investigations provide students with the opportunity to write and record their observations in their science notebooks. Students also take notes and organize information when reading articles. Selected examples Inv 1, Part 1, Steps 10-12; Inv 1, Part 3, Steps 10, 12 Inv 2, Part 1, Steps 8, 19, 23; Inv 2, Part 2, Steps 5, 8, 10, 12, 14, 17 Inv 2, Part 3, Steps 7, 9, 10, 12, 16 Inv 2, Part 1, Step 23; Inv 2, Part 2, Step 9 Inv 3, Part 1, Steps 12, 24, 30, 32, 35, 36, 39 Inv 3, Part 2, Steps 6, 8, 10, 16, 17 Inv 3, Part 3, Steps 4, 6, 14 Inv 4, Part 1, Steps 18, 27 Inv 4, Part 2, Steps 2, 14, 18, 19; Inv 4, Part 4, Steps 9-11
All investigations provide opportunities to use the *FOSS Science Resources* as a source from which to draw evidence to support their ideas (e.g., discussion questions at the end of the articles). Selected examples Inv 1, Part 2, Steps 20, 21 Inv 2, Part 4, Step 11 Inv 3, Part 3, Step 18 Inv 4, Part 1, Step 23; Inv 4, Part 3, Step 25 Inv 4, Part 4, Step 25 Inv 5, Part 1, Step 22	All investigations provide opportunities to use the *FOSS Science Resources* as a source from which to draw evidence to support their ideas (e.g., discussion questions at the end of the articles). Selected examples Inv 1, Part 2, Steps 5, 25, 26; Inv 1, Part 3, Steps 16, 17 Inv 1, Part 4, Steps 14, 15 Inv 2, Part 1, Step 23; Inv 2, Part 3, Steps 7, 10 Inv 3, Part 1, Steps 12, 32-36; Inv 3, Part 2, Step 10 Inv 3, Part 3, Steps 4, 6 Inv 4, Part 4, Step 11

FOSS and Common Core ELA — Grade 5

FOSS and Common Core ELA — Grade 5

SPEAKING AND LISTENING STANDARDS

	Grade 5 Standard	Mixtures and Solutions Module
Comprehension and Collaboration	1. Engage effectively in a range of collaborative discussions (one-on-one, in groups, and teacher-led) with diverse partners on *grade 5 topics and texts*, building on others' ideas and expressing their own clearly. a. Come to discussions prepared, having read or studied required material; explicitly draw on that preparation and other information known about the topic to explore ideas under discussion. b. Follow agreed-upon rules for discussions and carry out assigned roles. c. Pose and respond to specific questions by making comments that contribute to the discussion and elaborate on the remarks of others. d. Review the key ideas expressed and draw conclusions in light of information and knowledge gained from the discussions.	All investigations provide students ample opportunities to engage in a range of collaborative discussions. Students discuss before, during, and after the active investigation and during the Wrap-up/Warm-up section. Selected examples Inv 1, Part 1, Step 22; Part 2, Step 5 Inv 1, Part 3, Steps 6, 21, 23 Inv 2, Part 1, Steps 13, 24: Inv 2, Part 2, Step 13 Inv 3, Part 1, Step 21; Inv 3, Part 2, Step 21 Inv 3, Part 3, Steps 6, 14 Inv 4, Part 1, Steps 3, 11, 13, 25; Inv 4, Part 2, Step 6 Inv 5, Part 1, Steps 14, 22; Inv 5, Part 2, Steps 2, 20 Inv 5, Part 3, Steps 7, 10 Discuss articles in *FOSS Science Resources* in pairs, small groups, and whole class. Selected examples Inv 1, Part 2, Step 22, 25; Inv 1, Part 4, Step 16 Inv 3, Part 1, Step 19; Inv 3, Part 3, Step 20 Inv 4, Part 1, Step 24; Inv 4, Part 3, Step 6 Inv 5, Part 2, Step 17
	2. Summarize a written text read aloud or information presented in diverse media and formats, including visually, quantitatively, and orally.	Discuss articles read aloud in *FOSS Science Resources* and online activities. Selected examples Inv 1, Part 1, Step 21; Inv 1, Part 2, Steps 22, 25 Inv 2, Part 2, Step 10; Inv 2, Part 3, Step 10 Inv 3, Part 1, Step 19; Inv 3, Part 3, Step 8 Inv 4, Part 1, Steps 23, 24 Video discussions Inv 2, Part 3, Step 11; Inv 5, Part 3, Step 16
	3. Summarize the points a speaker makes and explain how each claim is supported by reasons and evidence.	All investigations provide students with opportunities to summarize discussion points and explain how claims are supported by reasons and evidence (e.g., students compare their responses to the focus question and the response sheets during the Wrap-up/Warm-up section). Other opportunities arise when students present information to their group or the whole class. Selected examples Inv 1, Part 2, Step 18 Inv 3, Part 3, Step 10 Inv 4, Part 4, Step 2; Inv 4, Part 4, Step 19

Earth and Sun Module	Living Systems Module
All investigations provide students ample opportunities to engage in a range of collaborative discussions. Students discuss before, during, and after the active investigation and during the Wrap-up/Warm-up section. Selected examples Inv 1, Part 1, Step 17; Inv 1, Part 2, Step 25 Inv 1, Part 3, Steps 1, 13, 28 Inv 2, Part 2, Step 15; Inv 2, Part 3, Step 1 Inv 2, Part 4, Step 32; Inv 2, Part 5, Steps 8, 14 Inv 3, Part 1, Step 20; Inv 3, Part 2, Step 11 Inv 4, Part 1, Steps 3, 5, 18, 27; Inv 4, Part 2, Steps 11, 15, 21 Inv 4, Part 3, Steps 9, 21, 29; Inv 4, Part 4, Steps 4, 5, 17, 20 Inv 5, Part 1, Steps 4, 6, 23; Inv 5, Part 3, Step 28 Discuss articles in *FOSS Science Resources* in pairs, small groups, and whole class. Selected examples Inv 1, Part 2, Steps 21, 22 Inv 2, Part 4, Steps 24, 27, 28 Inv 3, Part 2, Step 13 Inv 4, Part 1, Step 25; Inv 4, Part 3, Steps 24, 25, 27 Inv 5, Part 3, Step 26; Inv 5, Part 4, Step 18	All investigations provide students ample opportunities to engage in a range of collaborative discussions. Students discuss before, during, and after the active investigation and during the Wrap-up/Warm-up section. Selected examples Inv 1, Part 1, Step 12; Inv 1, Part 2, Steps 28, 30 Inv 1, Part 3, Step 21 Inv 2, Part 1, Steps 3-5, 28; Inv 2, Part 2, Steps 1, 4, 16, 19 Inv 3, Part 1, Step 41; Inv 3, Part 2, Step 18 Inv 4, Part 1, Step 32; Inv 4, Part 2, Step 21; Inv 4, Part 3, Step 13 Discuss articles in *FOSS Science Resources* in pairs, small groups, and whole class. Selected examples Inv 1, Part 2, Steps 6, 26; Inv 1, Part 4, Step 16 Inv 2, Part 1, Step 24; Inv 2, Part 2, Step 8, 9; Inv 2, Part 3, Steps 9, 10 Inv 3, Part 1, Step 12; Inv 3, Part 2, Steps 7, 9, 10
Discuss articles read aloud in *FOSS Science Resources* and online activities. Selected examples Inv 1, Part 3, Steps 24, 25, 29 Inv 2, Part 5, Step 17 Inv 3, Part 1, Steps 13, 14 Video discussions Inv 2, Part 4, Step 29; Inv 4, Part 3, Step 16 Inv 5, Part 1, Step 21; Inv 5, Part 3, Step 25	Discuss articles read aloud in *FOSS Science Resources* and online activities. Selected examples Inv 1, Part 2, Steps 6, 10, 24; Inv 1, Part 3, Steps 8, 16, 17 Video discussions Inv 1, Part 2, Step 3 Inv 2, Part 3, Steps 8, 11; Inv 2, Part 3, Step 9 Inv 3, Part 1, Steps 12, 32, 34; Inv 3, Part 2, Steps 5, 7 Inv 4, Part 3, Step 4; Inv 4, Part 4, Step 7
All investigations provide students with opportunities to summarize discussion points and explain how claims are supported by reasons and evidence (e.g., students compare their responses to the focus question and the response sheets during the Wrap-up/Warm-up section). Other opportunities arise when students present information to their group or the whole class. Selected examples Inv 4, Part 1, Step 26 Inv 5, Part 4, Step 18	All investigations provide students with opportunities to summarize discussion points and explain how claims are supported by reasons and evidence (e.g., students compare their responses to the focus question and the response sheets during the Wrap-up/Warm-up section). Other opportunities arise when students present information to their group or the whole class. Selected examples Inv 3, Part 3, Steps 7, 20

FOSS and Common Core ELA — Grade 5

SPEAKING AND LISTENING STANDARDS (CONT.)

Grade 5 Standard	Mixtures and Solutions Module
Presentation of Knowledge and Ideas	
4. Report on a topic or text or present an opinion, sequencing ideas logically and using appropriate facts and relevant, descriptive details to support main ideas or themes; speak clearly at an understandable pace.	All investigations provide students with the opportunity to report on what they know about science topics and to make claims based on their observations and experiences. In the Wrap-up/Warm-up section students share their answers to the focus question using evidence to support their ideas. Students report on what they learn from the text when discussing the articles in *FOSS Science Resources*. Selected examples Inv 1, Part 1, Steps 3, 7, 8, 15; Inv 1, Part 2, Steps 1, 8, 14; Inv 1, Part 3, Steps 10, 14, 19 Inv 1, Part 4, Steps 4, 7, 10, 13 Inv 2, Part 2, Step 6; Inv 2, Part 3, Steps 5, 8, 15 Inv 3, Part 1, Steps 4-6; Inv 3, Part 3, Step 5 Inv 3, Part 4, Step 13 Inv 4, Part 1, Step 15; Inv 4, Part 2, Step 9; Inv 4, Part 4, Steps 6, 13, 23, 27 Inv 5, Part 2, Steps 3, 16; Inv 5, Part 3, Steps 4, 6, 18, 19
5. Include multimedia components (e.g., graphics, sound) and visual displays in presentations when appropriate to enhance the development of main ideas or themes.	Inv 1, Part 2, Step 8 Inv 2, Part 1, Steps 16, 24; Inv 2, Part 2, Step 6 Inv 3, Part 3, Steps 5, 12 Inv 4, Part 4, Step 18 Inv 5, Part 2, Step 17
6. Adapt speech to a variety of contexts and tasks, using formal English when appropriate to task and situation. (See grade 5 Language standards 1 and 3 for specific expectations.)	All investigations provide students with situations in which they use either informal (small-group discussions) or formal discourse structures and procedures (whole-group sharing). Protocols and sentence frames are provided for students who need support. Selected examples Inv 1, Part 2, Step 23 Inv 2, Part 1, Step 27 Inv 4, Part 4, Step 19

Earth and Sun Module	Living Systems Module
All investigations provide students with the opportunity to report on what they know about science topics and to make claims based on their observations and experiences. In the Wrap-up/Warm-up section students share their answers to the focus question using evidence to support their ideas. Students report on what they learn from the text when discussing the articles in *FOSS Science Resources*. Selected examples Inv 1, Part 1, Steps 1, 7, 8, 12; Inv 1, Part 2, Step 13 Inv 1, Part 3, Steps 2, 3, 5, 7, 10, 20, 24; Inv 1, Language Extensions. Read Sun and shadow stories Inv 2, Part 1, Steps 9, 11; Inv 2, Part 4, Steps 18, 21, 30 Inv 2, Part 5, Steps 1, 2, 14, 16 Inv 3, Part 1, Steps 1, 8, 9, 12, 17; Inv 3, Part 3, Steps 2, 3, 5 Inv 3, Science Extensions. Explore careers in meteorology Inv 4, Part 1, Steps 16, 18, 23, 25, 27 Inv 4, Part 2, Steps 1, 2, 7, 10, 12, 15, 19, 24, 25; Inv 4, Part 4, Steps 1, 5, 22, 26, 27 Inv 5, Part 1, Steps 7, 14, 22; Inv 5, Part 2, Step 5; Inv 5, Part 3, Step 15, 17; Inv 5, Part 4, Steps 2, 4-6, 8, 13	All investigations provide students with the opportunity to report on what they know about science topics and to make claims based on their observations and experiences. In the Wrap-up/Warm-up section students share their answers to the focus question using evidence to support their ideas. Students report on what they learn from the text when discussing the articles in *FOSS Science Resources*. Selected examples Inv 1, Part 1, Steps 5, 7, 10; Inv 1, Part 2, Steps 25, 26 Inv 1, Part 3, Step 15; Inv 1, Part 4, Steps 2, 6, 9, 10, 14 Inv 2, Part 1, Steps 3, 11, 12, 19, 21, 22, 24 Inv 2, Part 2, Steps 9 -11, 13, 16 Inv 2, Part 3, Steps 9, 10, 12, 13, 15 Inv 3, Part 1, Step 12; Inv 3, Part 2, Steps 16-10, 14 Inv 3, Part 3, Step 6 Inv 4, Part 1, Steps 23, 26, 30 Inv 4, Part 2, Steps 11, 14, 15, 17,
Inv 1, Part 1, Step 17; Inv 1, Part 3, Steps 12, 20 Inv 2, Part 5, Steps 7, 9, 14, 21 Inv 3, Part 1, Steps 12, 13; Inv 3, Science Extensions. Draw atmospheric posters Inv 5, Part 1, Step 20	Inv 1, Part 2, Steps 21, 26 Inv 2, Part 2, Steps 9, 19; Inv 2, Part 3, Step 15 EL Note Inv 3, Part 2, Step 14; Inv 3, Science Extensions. Diagram an organ system
All investigations provide students with situations in which they use either informal (small-group discussions) or formal discourse structures and procedures (whole-group sharing). Protocols and sentence frames are provided for students who need support. Selected examples Inv 1, Part 3, Step 20 Inv 3, Part 1, Step 13 Inv 4, Part 2, Steps 23, 29 Inv 5, Part 1, Step 21	All investigations provide students with situations in which they use either informal (small-group discussions) or formal discourse structures and procedures (whole-group sharing). Protocols and sentence frames are provided for students who need support. Selected examples Inv 1, Part 2, Step 21; Inv 1, Part 4, Step 9 Inv 2, Part 2, Step 9 Inv 3, Part 1, Step 33; Inv 3, Part 2, Step 14

FOSS and Common Core ELA — Grade 5

LANGUAGE STANDARDS

	Grade 5 Standard	Mixtures and Solutions Module
Conventions of Standard English	1. Demonstrate command of the conventions of standard English grammar and usage when writing or speaking. a. Explain the function of conjunctions, prepositions, and interjections in general and their function in particular sentences. b. Form and use the perfect (e.g., *I had walked; I have walked; I will have walked*) verb tenses. c. Use verb tense to convey various times, sequences, states, and conditions. d. Recognize and correct inappropriate shifts in verb tense. e. Use correlative conjunctions (e.g., *either/or, neither/nor*).	All investigations provide opportunities for students to apply the conventions of English grammar when writing and speaking. Selected example Inv 1, Part 1, Step 14
	2. Demonstrate command of the conventions of standard English capitalization, punctuation, and spelling when writing. a. Use punctuation to separate items in a series. b. Use a comma to separate an introductory element from the rest of the sentence. c. Use a comma to set off the words *yes* and *no* (e.g., *Yes, thank you*), to set off a tag question from the rest of the sentence (e.g., *It's true, isn't it?*), and to indicate direct address (e.g., *Is that you, Steve?*). d. Use underlining, quotation marks, or italics to indicate titles of works. e. Spell grade-appropriate words correctly, consulting references as needed.	All investigations provide opportunities for students to demonstrate command of the conventions of standard English capitalization, punctuation, and spelling when writing in their science notebooks, response sheets, and I-Checks.
Knowledge of Language	3. Use knowledge of language and its conventions when writing, speaking, reading, or listening. a. Expand, combine, and reduce sentences for meaning, reader/listener interest, and style. b. Compare and contrast the varieties of English (e.g., dialects, registers) used in stories, dramas, or poems.	All investigations provide opportunities for students to use their knowledge of language and its conventions when writing in their science notebooks, discussing the investigation, and reading the articles in *FOSS Science Resources*. Selected examples Inv 3, Part 1, Step 21; Inv 3, Part 2, Step 21 Inv 3, Part 3, Step 13; Inv 3, Part 4, Step 8

Earth and Sun Module	**Living Systems Module**
All investigations provide opportunities for students to apply the conventions of English grammar when writing and speaking. Selected examples Inv 1, Part 1, Step 15 Inv 4, Part 3, Step 9 Inv 5, Part 3, Step 26	All investigations provide opportunities for students to apply the conventions of English grammar when writing and speaking. Selected examples Inv 1, Part 2, Step 28; Inv 1, Part 3, Step 12 Inv 2, Part 1, Step 26 Inv 3, Part 3, Step 14 Inv 4, Part 3, Step 8
All investigations provide opportunities for students to demonstrate command of the conventions of standard English capitalization, punctuation, and spelling when writing in their science notebooks, response sheets, and I-Checks.	All investigations provide opportunities for students to demonstrate command of the conventions of standard English capitalization, punctuation, and spelling when writing in their science notebooks, response sheets, and I-Checks.
All investigations provide opportunities for students to use their knowledge of language and its conventions when writing in their science notebooks, discussing the investigation, and reading the articles in *FOSS Science Resources*. Selected example Inv 2, Part 4, Step 20	All investigations provide opportunities for students to use their knowledge of language and its conventions when writing in their science notebooks, discussing the investigation, and reading the articles in *FOSS Science Resources*. Selected examples Inv 1, Part 4, Step 15

FOSS and Common Core ELA — Grade 5

LANGUAGE STANDARDS (CONT.)

	Grade 5 Standard	Mixtures and Solutions Module
Vocabulary Acquisition and Use	4. Determine or clarify the meaning of unknown and multiple-meaning words and phrases based on *grade 5 reading and content*, choosing flexibly from a range of strategies. a. Use context (e.g., cause/effect relationships and comparisons in text) as a clue to the meaning of a word or phrase. b. Use common, grade-appropriate Greek and Latin affixes and roots as clues to the meaning of a word (e.g., *photograph, photosynthesis*). c. Consult reference materials both print and digital, to find the pronunciation and determine or clarify the precise meaning of key words and phrases.	All investigations provide opportunities for students to determine or clarify the meaning of academic and science-specific words and phrases during class discussions and while reading and discussing articles in *FOSS Science Resources*. Selected examples Inv 2, Part 2, Step 10 Inv 3, Part 1, Step 18; Inv 3, Part 2, Step 19 Inv 3, Part 3, Step 9 Inv 4, Part 2, Step 10 Inv 5, Part 2, Step 16
	5. Demonstrate understanding of figurative language, word relationships, and nuances in word meanings. a. Interpret figurative language, including similes and metaphors, in context. b. Recognize and explain the meaning of common idioms, adages, and proverbs. c. Use the relationship between particular words (e.g., synonyms, antonyms, homographs) to better understand each of the words.	Students learn the word relationships (e.g., concept maps) and nuances of certain words that have a specific meaning in science, such as **mixture, separate, screen, filter, solution, dissolve, solvent, solute, mass, evaporation, crystal, criteria, constraints, extract, model, construct, siphon, melt, concentrated, dilute, volume, dense, saturated, soluble, gas, precipitate, reaction,** and **products.** Selected examples Inv 1, Part 1, Step 22; Inv 1, Part 2, Steps 16, 17 Inv 2, Part 1, Step 11 Inv 3, Part 4, Step 10; Inv 3, Language Extensions Inv 4, Part 3, Step 1; Inv 4, Part 4, Step 2 Inv 5, Part 2, Step 9; Inv 5, Language Extensions
	6. Acquire and use accurately grade-appropriate general academic and domain-specific words and phrases, including those that signal contrast, addition, and other logical relationships (e.g., *however, although, nevertheless, similarly, moreover, in addition*).	All investigations provide opportunities for students to acquire and use academic, and science-specific words and phrases. Science vocabulary words are in bold when they are first introduced to students in *FOSS Science Resources*. Students also review vocabulary in the Review vocabulary for each part and the Wrap-up section of each investigation. Selected examples Inv 1, Part 1, Steps 3, 9, 18; Inv 1, Part 2, Steps 15, 16; Inv 1, Part 3, Steps 1, 22; Inv 1, Part 4, Steps 13, 19 Inv 2, Part 2, Steps 8, 10, 13; Inv 2, Part 3, Steps 10, 12 Inv 3, Part 1, Steps 9, 13; Inv 3, Part 2, Steps 1, 15, 19; Inv 3, Part 4, Steps 1, 7, 8; Inv 3, Language Extensions. Inv 4, Part 1, Steps 10, 19; Inv 4, Part 2, Steps 10, 11 Inv 4, Part 4, Step 28 Inv 5, Part 1, Steps 1, 2, 14-16; Inv 5, Part 2, Steps 5, 9

Earth and Sun Module	**Living Systems Module**
All investigations provide opportunities for students to determine or clarify the meaning of academic and science-specific words and phrases during class discussions and while reading and discussing articles in *FOSS Science Resources*. Selected examples Inv 1, Part 2, Step 20 Inv 3, Part 1, Step 18	All investigations provide opportunities for students to determine or clarify the meaning of academic and science-specific words and phrases during class discussions and while reading and discussing articles in *FOSS Science Resources*. Selected examples Inv 1, Part 1, Step 11; Inv 1, Part 3, Steps 3, 16 Inv 2, Part 2, Step 9; Inv 2, Part 3, Step 5 Inv 3, Part 1, Step 12; Inv 3, Part 3, Steps 6, 7 Inv 4, Part 1, Step 26
Students learn the word relationships (e.g., concept maps) and nuances of certain words that have a specific meaning in science, such as **shadow, day, night, rotation, axis, sunrise, sunset, revolution, orbit, crescent, gibbous, lunar, waning, waxing, solar, planets, compress, pressure, matter, mass, atmosphere, forecasting, temperature, heat, transfer, conduction, dense, absorb, reflect, condensation, vapor, glaciers,** and **climate.** Selected examples Inv 2, Part 4, Step 16 Inv 3, Part 1, Step 15; Inv 3, Part 2, Step 11 Inv 4, Part 1, Steps 20, 21; Inv 4, Language Extension. List the effects of heat	Students learn the word relationships (e.g., concept maps) and nuances of certain words that have a specific meaning in science, such as **system, interact, geosphere, atmosphere, hydrosphere, biosphere, ecosystems, food chain, food web, producers, energy, algae, consumers, decomposers, recycle, marine, compost, nutrients, waste, sugar. fungus,, transpiration, vascular, heart, valves, vein, vital capacity, response, instinct, stimulus,** and **adaptation.** Inv 1, Part 1, Step 10; Inv 1, Part 2, Steps 7, 12-16; Inv 1, Part 3, Steps 2, 7; Inv 1, Part 4, Step 15 Inv 2, Part 1, Step 25 Inv 3, Part 1, Steps 12, 37
All investigations provide opportunities for students to acquire and use academic, and science-specific words and phrases. Science vocabulary words are in bold when they are first introduced to students in *FOSS Science Resources*. Students also review the vocabulary in the Review vocabulary section for each part and the Wrap-up section of each investigation. Selected examples Inv 1, Part 1, Step 14; Inv 1, Part 3, Steps 8, 9, 19, 21, 22 Inv 2, Part 4, Steps 14, 15, 16, 30; Inv 2, Part 5, Steps 3, 10 Inv 3, Part 1, Steps 10, 14, 15; Inv 3, Part 2, Steps 1, 4, 9 Inv 3, Part 3, Steps 3, 5, 6, 14 Inv 4, Part 1, Steps 1, 2, 5, 6, 17, 18-20; Inv 4, Part 2, Step 9 Inv 4, Part 3, Steps 1, 19; Inv 4, Part 4, Steps 3, 21, 23 Inv 5, Steps 1, 5, 16, 17; Inv 5, Part 2, Steps 6, 7; Inv 5, Part 3, Steps 18, 27; Inv 5, Part 4, Steps 3, 11, 15	All investigations provide opportunities for students to acquire and use academic, and science-specific words and phrases. Science vocabulary words are in bold when they are first introduced in *FOSS Science Resources*. Students also review the vocabulary in the Review section for each part and the Wrap-up section of each investigation. Selected examples Inv 1, Part 1, Steps 1, 4, 6, 8; Inv 1, Part 2, Steps 3, 7, 9, 12-16, 27; Inv 1, Part 3, Steps 1, 3, 4, 11; Inv 1, Part 4, Steps 2, 4, 13 Inv 2, Part 1, Steps 2, 3, 15, 21, 25; Inv 2, Part 3, Step 14 Inv 3, Part 1, Steps 8, 9, 29, 37; Inv 3, Part 2, Steps 3, 5, 15; Inv 3, Part 3, Steps 3, 6, Inv 4, Part 1, Steps 1, 3, 4, 23, 25; Inv 4, Part 2, Steps 12, Inv 4, Part 3, Steps 1, 6

FOSS and Common Core ELA — Grade 5

FOSS and Common Core ELA — Grade 5

FOSS and Common Core Math – Grade 5

FOSS and Common Core Math — Grade 5

Contents

Introduction 1

Operations and Algebraic
Thinking for Grades 4–5 4

Number and Operations in
Base Ten for Grade 5................6

Number and Operations—
Fractions for Grades 4–5 8

Measurement and Data
for Grades 4–5........................10

Geometry for Grade 5............14

INTRODUCTION

The adoption of the Common Core State Standards for Mathematics calls for shifts in focus, coherence, and rigor. The teaching of the standards should be focused on the important content, coherent from one grade level to the next, and rigorous in requiring conceptual understanding, fluency, and application. Within this area of application, FOSS provides fertile ground for the use of mathematics.

The FOSS Program integrates mathematics with science in two ways throughout the grade 5 modules. In active investigations, students apply mathematics during data gathering and analysis. In addition, the Interdisciplinary Extensions at the end of each investigation usually include a math problem of the week. These problems enhance the science learning by providing hypothetical data for students to analyze or in some way relate to the context of the investigation. The notes explain for the teacher the problem and describe how students might approach its solution. The problems are prepared for distribution to students on duplication masters in the Teacher Masters chapter of *Teacher Resources*.

This chapter gives an overview of how FOSS addresses the Common Core State Standards for Mathematics through science. It also points out specific instances in which students exercise those skills during science instruction.

Full Option Science System   *Copyright © The Regents of the University of California*

FOSS and Common Core Math — Grade 5

Mathematical Practices

Mathematical practices consist of eight processes and proficiencies that are important for all students.

1. Make sense of problems and persevere in solving them.
2. Reason abstractly and quantitatively.
3. Construct viable arguments and critique the reasoning of others.
4. Model with mathematics.
5. Use appropriate tools strategically.
6. Attend to precision.
7. Look for and make use of structure.
8. Look for and express regularity in repeated reasoning.

Within the context of science, students use some of these mathematical practices on a regular basis. According to *Next Generation Science Standards* (volume 2, appendix L, p. 138).

The three CCSSM practice standards most directly relevant to science are:

- MP.2. Reason abstractly and quantitatively.
- MP.4. Model with mathematics.
- MP.5. Use appropriate tools strategically.

When students reason abstractly and quantitatively and model with mathematics, they are using math in context. They work with symbols and their meanings and represent and solve word problems. Students choose and correctly use the available tools to collect data and solve problems. In the grade 5 modules, students engage with these three practices during the active investigation and by completing the problems at the end of each investigation. Here are some examples.

In solving the Math Problem of the week for Investigation 1 of the **Living Systems Module**, students model a particular worm bin given variables that impact the composting of lunch scraps. In order to solve this, students reason quantitatively and use mathematics in the context of science. They can use drawings to determine the area of the bin, visually represent the amount of food worms eat, and the number of scraps students produce. Students can then reason quantitatively as they multiply or divide to answer questions.

In the **Mixtures and Solutions Module**, students are presented with a challenge to redistribute unequal amounts of liquid so each container has the same amount. The problem appears simple, but students can approach it in different ways. Students can combine the liquids in different containers to make full cups and then redistribute or they can reason abstractly to pour liquid from almost full cups to almost empty cups until they all have the same. Story problems provide opportunities for students to utilize tools to determine the solutions to two-step problems.

In the **Earth and Sun Module**, students are asked to use tools to solve a problem about the orbits of different fictional moons. This requires students to create a table to determine when the moons with different orbits will be lined up. Recording the orbit data in a table reveals common multiples which can then be used to answer various questions about the number of orbits and future alignment between the moons.

Mathematical Content

The mathematical content in fifth grade is organized around five concepts.

- Operations and algebraic thinking
- Number and operations in base 10
- Number and operations—fractions
- Measurement and data
- Geometry

The following pages have a table that identifies the opportunities to engage students in developing these mathematical concepts as well as those learned in grade 4. It lists the math content for some fourth and fifth grades and points out relevant opportunities in the three FOSS modules for grade 5.

FOSS and Common Core Math — Grade 5

OPERATIONS AND ALGEBRAIC THINKING FOR GRADES 4–5

	Standard	Mixtures and Solutions Module
Grade 4	**Use the four operations with whole numbers to solve problems.**	
	2. Multiply or divide to solve word problems involving multiplicative comparison, e.g., by using drawings and equations with a symbol for the unknown number to represent the problem, distinguishing multiplicative comparison from additive comparison.	Inv 3, Extension, Calculate drink cost
Grade 5	**Write and interpret numerical expressions.**	
	1. Use parentheses, brackets, or braces in numerical expressions, and evaluate expressions with these symbols.	

Common Core State Standards for Mathematics (National Governors Association Center for Best Practices and Council of Chief State School Officers, 2010).

Earth and Sun Module	Living Systems Module
	Inv 2, Problem of the week

FOSS and Common Core Math — Grade 5

NUMBER AND OPERATIONS IN BASE TEN FOR GRADE 5

Standard	Mixtures and Solutions Module
Perform operations with multi-digit whole numbers and with decimals to hundredths.	
5. Fluently multiply multi-digit whole numbers using the standard algorithm.	
6. Find whole-number quotients of whole numbers with up to four-digit dividends and two-digit divisors, using strategies based on place value, the properties of operations, and/or the relationship between multiplication and division. Illustrate and explain the calculation by using equations, rectangular arrays, and/or area models.	Inv 4, Problem of the week
7. Add, subtract, multiply, and divide decimals to hundredths, using concrete models or drawings and strategies based on place value, properties of operations, and/or the relationship between addition and subtraction; relate the strategy to a written method and explain the reasoning used.	Inv 5, Problem of the week

Earth and Sun Module	Living Systems Module
Inv 2, Part 2, Step 4, Provide size and distance data Inv 2, Part 2, Step 8, Introduce Sun model	
	Inv 1, Problem of the week Inv 2, Problem of the week

FOSS and Common Core Math — Grade 5

NUMBER AND OPERATIONS—FRACTIONS FOR GRADES 4–5

	Standard	Mixtures and Solutions Module
Grade 4	**Extend understanding of fraction equivalence and ordering.**	
	1. Explain why a fraction *a/b* is equivalent to a fraction *(n × a)/(n × b)* by using visual fraction models, with attention to how the number and size of the parts differ even though the two fractions themselves are the same size. Use this principle to recognize and generate equivalent fractions.	Inv 3, Part 2, Step 17, Assess progress: response sheet
Grade 5	**Use equivalent fractions as a strategy to add and subtract fractions.**	
	1. Add and subtract fractions with unlike denominators (including mixed numbers) by replacing given fractions with equivalent fractions in such a way as to produce an equivalent sum or difference of fractions with like denominators. *For example, 2/3 + 5/4 = 8/12 + 15/12 = 23/12. (In general, a/b + c/d = (ad + bc)/bd.)*	Inv 3, Problem of the week

Earth and Sun Module	Living Systems Module

FOSS and Common Core Math — Grade 5

MEASUREMENT AND DATA FOR GRADES 4–5

Standard	Mixtures and Solutions Module
Solve problems involving measurement and conversion of measurements from a larger unit to a smaller unit.	
3. Apply the area and perimeter formulas for rectangles in real world and mathematical problems. *For example, find the width of a rectangular room given the area of the flooring and the length, by viewing the area formula as a multiplication equation with an unknown factor.*	Inv 2, Problem of the week Inv 2, Extension, Draw blueprints
Geometric measurement: understand concepts of angle and measure	
5. Recognize angles as geometric shapes that are formed wherever two rays share a common endpoint, and understand concepts of angle measurement: a. An angle is measured with reference to a circle with its center at the common endpoint of the rays, by considering the fraction of the circular arc between the points where the two rays intersect the circle. An angle that turns through 1/360 of a circle is called a "one-degree angle," and can be used to measure angles. b. An angle that turns through *n* one-degree angles is said to have an angle measure of *n* degrees.	
6. Measure angles in whole-number degrees using a protractor. Sketch angles of specified measure.	

(Grade 4)

Earth and Sun Module	Living Systems Module
Inv 1, Part 2, Step 16, Replicate trackings	
Inv 1, Extension, Use circles and degrees	

FOSS and Common Core Math — Grade 5

MEASUREMENT AND DATA FOR GRADES 4–5 (CONT.

Standard	Mixtures and Solutions Module
Convert like measurement units within a given measurement system.	
1. Convert among different-sized standard measurement units within a given measurement system (e.g., convert 5 cm to 0.05 m), and use these conversions in solving multi-step, real world problems.	
Represent and interpret data.	
2. Make a line plot to display a data set of measurements in fractions of a unit (1/2, 1/4, 1/8). Use operations on fractions for this grade to solve problems involving information presented in line plots. *For example, given different measurements of liquid in identical beakers, find the amount of liquid each beaker would contain if the total amount in all the beakers were redistributed equally.*	Inv 3, Problem of the week
Geometric measurement: understand concepts of volume and relate volume to multiplication and to addition.	
5. Relate volume to the operations of multiplication and addition and solve real world and mathematical problems involving volume. a. Find the volume of a right rectangular prism with whole-number side lengths by packing it with unit cubes, and show that the volume is the same as would be found by multiplying the edge lengths, equivalently by multiplying the height by the area of the base. Represent threefold whole-number products as volumes, e.g., to represent the associative property of multiplication. b. Apply the formulas $V = l \times w \times h$ and $V = b \times h$ for rectangular prisms to find volumes of right rectangular prisms with whole-number edge lengths in the context of solving real world and mathematical problems. c. Recognize volume as additive. Find volumes of solid figures composed of two non-overlapping right rectangular prisms by adding the volumes of the non-overlapping parts, applying this technique to solve real world problems.	

Earth and Sun Module	Living Systems Module
	Inv 1, Problem of the week
Inv 3, Problem of the week Inv 3, Extension, Use spreadsheet to look at weather data	
Inv 4, Problem of the week	

FOSS and Common Core Math — Grade 5

GEOMETRY FOR GRADE 5

Standard	Mixtures and Solutions Module
Graph points on the coordinate plane to solve real-world and mathematical problems.	
1. Use a pair of perpendicular number lines, called axes, to define a coordinate system, with the intersection of the lines (the origin) arranged to coincide with the 0 on each line and a given point in the plane located by using an ordered pair of numbers, called its coordinates. Understand that the first number indicates how far to travel from the origin in the direction of one axis, and the second number indicates how far to travel in the direction of the second axis, with the convention that the names of the two axes and the coordinates correspond (e.g., *x*-axis and *x*-coordinate, *y*-axis and *y*-coordinate).	
2. Represent real world and mathematical problems by graphing points in the first quadrant of the coordinate plane, and interpret coordinate values of points in the context of the situation.	

Earth and Sun Module	Living Systems Module
Inv 1, Problem of the week	
Inv 1, Extension, Match bar graphs to shadow scenarios Inv 3, Part 2, Step 12, Demonstrate weather monitoring and recording Inv 3, Problem of the week Inv 3, Extension, Use spreadsheet to look at weather data Inv 4, Part 1, Step 16, Graph and analyze the results Inv 4, Part 4, Step 19, Graph the data Inv 4, Extension, How does distance from the Sun affect a planet's temperature? Inv 5, Problem of the week	

FOSS and Common Core Math — Grade 5

Taking FOSS Outdoors

Taking FOSS Outdoors

If we want children to flourish, to become truly empowered, then let us allow them to love the earth before we ask them to save it.

David Sobel, *Beyond Ecophobia*

Contents
Introduction 1
What Does FOSS Look Like Outdoors? 2
Goals and Objectives 3
Managing Space 4
Managing Time 8
Managing Materials 10
Managing Students 13
Teaching Strategies 20
Flow of Outdoor Activities 22
Extending beyond FOSS Outdoor Activities 23
Elementary-Level Environmental Education 25
References 27
Acknowledgments 28

INTRODUCTION

During its first 20 years, FOSS focused on classroom science. The goal was to develop a scientifically literate population with an ever-growing knowledge of the natural world and the interactions and organizational models that govern and explain it. In recent years, it has become clear that we have a larger responsibility to the students we touch with our program. We have to extend classroom learning into the field to bring the science concepts and principles to life. In the process of validating classroom learning among the schoolyard trees and shrubs, down in the weeds on the asphalt, and in the sky overhead, students will develop a relationship with nature. It is our relationship with natural systems that allows us to care deeply for these systems. In order for students in our schools today to save Earth, and save it they must, they first have to feel the pulse, smell the breath, and hear the music of nature. So pack up your explorer's kit, throw open the door, and join us. We're taking FOSS outdoors.

Taking FOSS Outdoors

WHAT DOES FOSS LOOK LIKE OUTDOORS?

Visualize taking FOSS outdoors: Students exit the classroom in an orderly fashion, their direction and purpose undeterred by the joyful sounds of other students at recess. With focused enthusiasm, the band of young scientists moves toward the edge of the schoolyard. Each student is carrying something, maybe a clipboard for recording, a container for collecting, or a hand lens for observing. Students reach their destination and quickly form a sharing circle. After a brief orientation, students disperse and begin searching the tall grass along the chain-link fence. All are independently recording in their science notebooks, and all are on task. The teacher moves about with intention, speaking to a few students at a time. After several more minutes of this work, the teacher rings a chime. Students freeze, raise one arm, and look at her. She rings the chime again. Students leave their materials in their spots and re-form their sharing circle with their teacher for discussion or additional instructions.

This scenario could be anywhere in the country with a regular classroom teacher using any of the FOSS modules. Taking FOSS outdoors is a natural extension of the classroom work. It looks and feels a lot like standard FOSS activities. Many of the routines you use inside the classroom can be implemented outdoors as well. Success, however, does depend on a few specialized skills and specific preparation to maximize outdoor teaching efficiency.

Expect the enthusiasm, participation, engagement, group discussions, and effort on notebook entries to be heightened during and after an outdoor experience. Even the simplest outdoor activities create a surge of positive energy. It is difficult to determine whether the enthusiasm and commitment students exhibit when doing FOSS outdoors comes from exercising content they already know or from just being outdoors. Students can be a bit louder and more excited when they are learning outdoors, and the space allows for this expansion in energy level, which benefits some students immensely.

GOALS AND OBJECTIVES

The three program goals set down 20 years ago still serve FOSS well. They are: (1) scientific literacy for all students, (2) instructional efficiency and support for teachers, and (3) systemic reform.

The march into the schoolyard has three objectives that relate to the goal for students. First, the outdoor activities **continue and extend the learning** that starts in the classroom. The outdoor activities provide more experience with the content and additional opportunities to practice skills and techniques developed in the classroom.

Second, venturing out provides opportunities for students to discover **applications and examples of classroom content and concepts**. The classroom activities work well for developing sound conceptual science knowledge. That knowledge, however, is constrained by the context in which the concepts are taught. For students to take the next level of ownership of that knowledge, they need to see how it applies and generalizes in the broader context of the world. Leaving the classroom context with a head full of new ideas and new tools for observation enriches the learning.

The third objective is to **connect students with nature**. On the boundaries of the planned, structured experiences are the intangibles that may spark a new relationship with natural systems. It may start with a multisensory experience in the native environment—wind, cold, sunshine, plants, insects, on and on—and advance to an awareness of the diversity of resources surrounding the school. It might evolve into a consciousness of place, followed by a flood of questions about the structure, organization, and operation of the schoolyard ecosystem. When students bond with nature, they have accepted a precious gift, and we have accomplished something important.

Taking FOSS Outdoors

Taking FOSS Outdoors

MANAGING SPACE

FOSS outdoor activities are designed to be successful in a diversity of schoolyards. Some schoolyards are covered in asphalt, while others have been turned into well-designed outdoor learning environments. Some include large, grassy areas without trees, and others are covered with mulch. One outdoor space may be circled by a variety of mature trees; the next may have recently planted maples and pines scattered about. The space may reflect thoughtful attention or neglect. Nevertheless, FOSS believes that bringing students into the fresh air under a changing sky, into the available outdoor space, will awaken their well-being and stimulate their understanding of science concepts.

Choosing Outdoor Spaces

Whether your school's landscape is wild, manicured, or asphalt, there are more options for outdoor learning spaces than might initially meet the eye. This section will help you choose the best spaces near your school for the FOSS outdoor activity.

Before choosing your outdoor study areas, get to know your outdoor spaces. Look closely at all areas surrounding the school building—even places that students do not normally go. Many seemingly uninteresting monoculture fields are flourishing with a diversity of different grass species and other small flowering plants. Consider the pile of leaves that blew into a corner of your schoolyard; a crack in the concrete; or the ragged, weedy edge of the field where the lawn mower doesn't reach. These are places that provide small animals with what they need—food, shelter, water, and space. Transition zones where vegetation changes from shrubbery to lawn or garden to field can present interesting study sites. As you ponder the learning possibilities in and around your schoolyard, consider these characteristics.

Accessibility. You should be able to walk from your classroom to your outdoor site in 2–5 minutes. Sites farther than 10 minutes away can be considered for special outings, but are not realistic for frequent access. Check out physical access if you have students whose mobility requires consideration. Be aware of slippery surfaces from water or ice, and caution students to be careful.

Purpose. Determine the space needs of the activity. Some activities will require open space, such as a field or blacktop. Other activities work better if students have a more diverse landscape with varying environmental conditions (such as temperature, light intensity, wind). Some activities require a variety of human-made materials to measure or test for certain properties (such as magnets to test for magnetism). Different areas will serve different needs.

Size. The space should be large enough for the class to work comfortably but small enough for you to supervise all students easily. You always need to be able to see all your students, and your students need to be close enough to hear you and your attention signal.

Boundaries. For any space you intend to use, make sure you have clearly defined the boundaries before heading outdoors with your students. Ideally, the landscape will be helpful. For example, stay between the sidewalk and the tree line. If natural markers are not present, you may need to bring along traffic cones or their equivalent to define limits. In general, consider if there are any hazards, such as dangerous debris, poisonous plants, or traffic.

Fostering and Maintaining Diversity

For life science and earth science studies, ideally you want your site to have a variety of living and dead plant matter and a range of environmental conditions. Survey your site to see if it includes places that have been left unmanaged. Even a small wild zone along a fence or behind the maintenance area or an adjacent field can be a valuable resource. It is important for students to see that living things carry on, even in the city, without human assistance.

Enhancing your schoolyard. You may be able to secure a small section of the schoolyard from the school custodian, allowing it to grow wild to compare to the managed school grounds. Consult with your custodian and principal to see if this can be arranged.

Another way to enhance biodiversity is to encourage decomposition by letting fallen leaves and/or lawn clippings to remain on an area of soil over the winter. This gives worms and other decomposers something to eat, which, in turn, provides food for everything else. Make sure all necessary parties are aware of your intentions to leave an area untouched. If you find that you need administrative permission, consider ways to contain and mark the unkempt (but not unloved) area so that it clearly represents an intentional project.

Taking FOSS Outdoors

Taking FOSS Outdoors

Tread lightly. Your schoolyard study areas will potentially experience some user impact. It is important to teach students to minimize their environmental footprint. Otherwise, the living things they disturb might seek a safer place to live. Unless the class is intentionally collecting specimens, nothing natural should be picked up or removed from the area. This is a good opportunity to introduce the "leave no trace" philosophy, which, in an effort to preserve an area for recreation, encourages us to leave natural objects as we find them.

At some schools, the outdoor space is used by so many classrooms that a system is needed to schedule outdoor activities. A sign-up sheet can be used to reserve outdoor spaces just as is done to reserve other school resources. Check the site the morning before taking the class outdoors to make sure the area is ready for students to investigate.

Weather

Weather can present great challenges and exceptional experiences. Inclement weather can provide an excellent opportunity to study environmental concepts: water drainage, wind impact, plant and animal survival adaptations. (There is nothing like being out in a snowstorm to appreciate the value of insulation!) Making extra preparations to study out in the elements has value. If the activity can be undertaken with some assurance of success, try to make it work. Over time, students acclimate to all sorts of weather and will actually look forward to the challenge of going out in difficult weather.

Clothing. The right gear at the right time can make all the difference. Baseball caps stored at school can work well in a light rain and are often essential as sun protection in warmer climates. Baseball caps in a light rain are especially helpful for students who wear glasses. If possible, invest in a set of rain ponchos to make it possible to go out in wet weather. Large trash bags can make very effective, low-cost ponchos. Of course, you will want to model this elegant attire. Communicate regularly with students and family members about upcoming outdoor experiences.

Wind. A stiff breeze can fling your materials into disarray or send notebooks flying. If you anticipate wind, discuss ways to keep materials from blowing away (such as using natural paperweights or taping down nonliving specimens). If there is a protected area where you and your class can take shelter briefly, the activity can continue. You may have to chase down a couple of notebook sheets before students become accustomed to securing papers and other light materials.

Safety and Comfort

Be prepared for the unexpected. Insect stings (ants, bees, wasps, mosquitoes) can be alarmingly painful for young students, particularly if they have not been stung before. Although extremely unlikely in a schoolyard, have a plan developed with students in advance as to how to retreat with purpose if someone accidentally disturbs a nest. You should already know who is allergic and who has never been stung before.

Skin-irritating plants (poison oak, poison ivy, poison sumac, nettles) can certainly put a damper on a field trip. Take a moment, and get to know your local irritants and toxic plants. The rule "leaves of three, let it be" works only for poison ivy and poison oak. Poison sumac has 7–13 leaves on a branch. Stinging nettle feels much like being stung by a jellyfish and can be very frightening for students who have never experienced it. Often, the irritation subsides within a few minutes; do not treat rashes with bleach or rubbing alcohol.

Lyme disease is a treatable bacterial infection, carried by deer ticks. It is present throughout the country, but is particularly present in eastern states. It is possible to get sick without finding a tick bite. If you or your students experience flulike symptoms that are severe enough to see a doctor, make sure that doctor is aware of any outdoor exposure.

If you are out and about in tick country, tuck pant legs into socks and take a few minutes at the end of the trip to pair up and look for obvious ticks on clothing and on the neck and shoulders of a partner.

Taking FOSS Outdoors

MANAGING TIME

When to Teach

When you start a new module, anticipate when you might want to go outdoors, and schedule the time. The At a Glance chart in each investigation can help with this planning.

Time of year. If possible, plan the time of year when you will teach particular modules. In the northern tier, life science and earth science modules would be best in the fall or spring. In the southern tier, it might be best to teach life science modules in the winter when it is not uncomfortably hot during the day. Good times to coordinate your outdoor activities with the school calendar include minimum days or other disruptions to the regular schedule, days just before or after school vacations, and days following district testing.

Time of day. Consider the time of day you teach your activities. Established schedules are often difficult to alter, but you might find it advantageous to do so. If you do a lot of seat work in the mornings, you may want to break the routine occasionally with an outdoor activity. Students will return to the classroom refreshed and ready to focus on the next seated activity you have planned.

If you live in a climate where it gets really hot during the school day, you might want to teach outdoors early in the day. Conversely, if you live in a cold climate, you might want to do your winter outdoor activities midday. If you're looking for wildlife (birds, insects, mammals), the best time to go outdoors might be in the morning.

If you plan to use a part of the schoolyard that is heavily populated at predictable times during the day (lunch, physical education), plan to venture out at a time when other activities are minimal.

Stay flexible. If you are studying the **Water and Climate Module**, for example, be prepared to dash out if it rains or snows. One of the delights of outdoor education is going out when nature is putting on a show. Inquiring minds rush out for the experience when timid observers retreat.

Specific times. Some activities require a sunny day. Measuring shadows, solar water heaters, and solar cell investigations require sunshine. It can be tricky to move on without completing specific observations or experiments. Be creative. You may need to proceed with the module and return to the sunny-day activity when the Sun finally comes out.

Instructional Time

An outdoor activity might require 15 minutes, or it might require an hour. Only part of the time budgeted for outdoor learning is actually spent interacting with the schoolyard terrain, plants, and animals. The rest is management.

Travel time. It will take perhaps 10 minutes from the announcement that it is time to decamp for the schoolyard and the time you arrive there. It will take several minutes to describe and distribute materials, get the appropriate clothing, line up, and travel in an orderly fashion to the designated location. Travel back to the classroom will take another 3–4 minutes.

Instructions. Outdoors, students form a sharing circle. It will take 2–4 minutes to review rules, set the boundaries for the activity, describe the challenge, and distribute materials.

Investigation time. Students break into pairs or groups to engage in the outdoor investigation. This might be as short as 8–10 minutes or as long as 30–40 minutes.

Wrap-up. Students return to the sharing circle to share and discuss their discoveries for several minutes.

Classroom follow-up. Frequently, students bring artifacts back to class to display in a classroom museum or to set up for further observation.

▶ **SAFETY NOTE**
Students should not disturb or collect live organisms in their natural habitats.

Some outdoor activities call for more flexible allocations of time. An activity may call for setup early in the day with periodic monitoring or measuring throughout the day.

Taking FOSS Outdoors

Taking FOSS Outdoors

MANAGING MATERIALS

When students step onto the schoolyard, they are field scientists. In the field, there is no lab bench where investigations can be set up, and there is no ready supply of materials. The field equipment must be minimal, portable, and durable so that it can be easily and safely transported from the classroom and back.

Field Equipment

A student's outdoor bag will contain the specific materials needed for the activity of the day as well as some core necessities, such as a hand lens and a writing tool.

Student outdoor bags might contain these basics.

- Pencils/pens
- Hand lens on brightly colored string or yarn
- Colored pencils or crayons
- Measuring tape
- Vials with caps
- Clipboard or notebook
- Seat pad

Note that pens and pencils each have drawbacks: pencil points break, and pen ink freezes in extremely cold weather. Seat pads can simply be several sheets of newspaper covered with a plastic bag.

Your basic teacher's outdoor equipment bag will include a few backup student materials and some items for helping with management.

- Extra pencils, pens, hand lenses, vials, and cups
- Attention signal (chime, whistle, or cowbell)
- Tissues and paper towels
- Basic first-aid kit (adhesive bandages)
- Phone (if leaving the school grounds)
- Student class list (particularly if you teach more than one class) with appropriate student health information and/or permission slips if away from school.

Transporting Materials

Getting materials to and from the outdoor site is a shared responsibility. Students will carry their personal equipment, and class materials can be distributed among students or tackled as a teacher task. Students always carry something to the outdoor site, even when it would be easier for you to carry everything. This reminds students that they are heading out for science, not recess. A hand lens serves as such a token.

Some teachers prefer to have students carry only their clipboards or notebooks and pencils, while the teachers carry all the field equipment in a canvas shopping bag or milk crate to the outdoor home base. Other teachers use a wagon or wheelie crate to transport the equipment. After teaching a few outdoor activities, you will discover what works best for you. Students will get excited when they see you preparing your transport system for an outdoor activity.

Water. Water is often used during outdoor activities. If you are lucky, there will be a tap near your study site. More likely, you will carry water from the school building. Recycled plastic jugs with screw caps and smaller bottles with screw caps are good vessels.

At times, you will want open containers of water, such as buckets or basins (for washing rocks, cleaning containers, and so on). Half-filled buckets can be carried a short distance, but basins should be carried empty and filled from jugs.

You rarely have to bring water back inside. Recycle leftover water by watering schoolyard plants. Make this practice overt to help students develop respect for this vital natural resource.

Creating Outdoor Tools

A sturdy writing surface is essential for science in the schoolyard. A bound notebook (composition book) is excellent. A serviceable clipboard can be made from a piece of cardboard and a binder clip. Use a paper cutter to cut sturdy cardboard slightly larger than a sheet of notebook paper. Place a medium-size binder clip at the top and a large elastic band around the bottom (to keep the paper from flapping up). Tie a pencil on a brightly colored string to the binder clip.

Taking FOSS Outdoors

Taking FOSS Outdoors

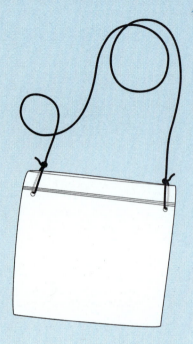

A group writing surface is important sometimes. You can use blue painter's tape to attach a sheet of chart paper temporarily to a wall or clip it onto a chain-link fence with binder clips or clothespins. On windy days, attach all four corners.

A small pack can serve as a hands-free means for students to tote their equipment. Little backpacks are excellent, but a serviceable low-cost satchel can be crafted from a large plastic bag and string. Purchase enough gallon-size zip bags for your class. Punch two holes just *under* the ends of the zipper. (This reduces tearing.) Cut the strings about 1 meter long. Tie sturdy knots that will not come undone. Store the string inside the bag after use to prevent tangling with other bags.

Hand lenses may disappear when students place them on the ground to perform a task. Run bright-colored string or yarn through holes in the lenses for students to wear. If your hand lenses do not already have holes, see if you can get holes drilled through the handles.

MANAGING STUDENTS

Going outdoors regularly is the best way to develop a productive and joyful working relationship with students in the outdoors. When students realize that going outdoors to learn is not a special event but, rather, a science event that will occur routinely, you may be surprised at how quickly they adapt to their expanded, enriched classroom.

Before the First Outing

It is always important to let the school administration know that you and your students will periodically be out of the classroom. If you are planning to leave the school grounds, remember to file a flight plan describing your itinerary, anticipated time of return, and contact partners.

At the beginning of the school year, send a letter home to families, letting them know that learning will extend to the schoolyard and, possibly, beyond. It may be possible to have a signed permission slip for impromptu walking field trips outside the schoolyard. Have families put their contact information and specific student health information on the permission slip. Photocopy these, and have one set of copies in the office and another set in a zip bag in your outdoor equipment bag for emergencies.

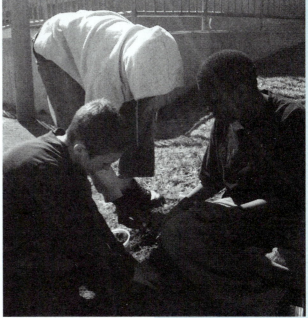

Tell students at the beginning of the year that they will be going outdoors often during science class. Remember to let them know a day in advance that they will be going outdoors. Let them know what it means to dress appropriately. This is especially important in the cold or stormy season when students will need proper clothing for safety and comfort. Your class can go out in any weather if students are dressed appropriately. A consistent system of reminders and clothing preparation will train students to be prepared.

Safety rules. Creating consistent, considerate rules of engagement is important. Learning is enhanced, and behavior problems are largely averted by routines that students participate in and understand.

Have a discussion about what students think constitutes proper preparation and behavior for leaving the classroom to study outdoors. (This discussion may be most productive *after* an initial orientation

Taking FOSS Outdoors

Taking FOSS Outdoors

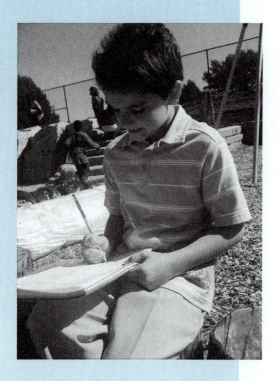

excursion to survey the schoolyard resources.) Have students generate a list of behaviors that they can adopt and respect. You may want to generate a second list of behaviors that you will agree to as leader of the adventure. Introduce as much formality into the process as you deem important. Develop the idea of a contract that all members of the class sign. Post the contract in your classroom and reference the *Outdoor Safety* poster. Here are the behaviors and rules that should appear on the list.

- Walk quickly and quietly outdoors.
- Outdoor science is not recess.
- Listen to the teacher's instructions.
- Freeze when the teacher rings the bell.
- Stay inside the boundaries.
- Don't make noise near the classrooms.
- Don't injure plants and animals in any way.
- Leave the outdoor environment the way you found it. Never release living organisms into the local environment unless they were collected there.

First Outing

Your first trip to the schoolyard may be a bit chaotic. Students may be distracted by other activities going on, and they may lapse into recess mode. A few precautions will minimize disruptive behaviors.

The path of egress. Determine which doors are available to access your outdoor site. Make sure you follow your school's policy for using auxiliary doors during the school day. Do they need to be closed at all times, or can you prop one open? Are they locked from the outside? You may be able to access keys in order to reduce travel time. When possible, avoid using the door you would normally use for recess. Students have a different mind-set when they walk onto the schoolyard through the "outdoor classroom" door.

Orientation activity. Consider an orientation activity for your first outing. The stated goal might be a site tour to inventory the resources on hand. Your primary agenda, however, is to dissipate the energy generated by the novelty of leaving the classroom during class time. Focus on preparing your transition to the outdoors, moving out in a purposeful and orderly fashion and arriving at your predetermined "home base," a destination that you will always go to initially when you leave the classroom. Form a sharing circle, a process you will use time and again. Tour the schoolyard, proceeding as a whole group, then ask students to walk as individuals for a few minutes, then with a partner for a minute, and finally with a group of four. This gives students a brief experience with each of the four ways they will be organized for various outdoor activities. End with another sharing circle and an orderly return to the classroom.

Challenging students. Sometimes the class will be inattentive and unresponsive. At such times, it is appropriate to direct students back to the classroom. Breaking the contract has consequences, and students need to understand that the opportunity to learn in nature is a privilege. They will remember that day.

In rare instances, you may have an individual student who is regularly not able to comply. Interestingly, this is probably not the student who you anticipated would have difficulty. Often, students who have difficulty with attention and performance in traditional classroom seat work shine and take leadership outdoors. In the case of the noncompliant student, it may be necessary to ask him or her to take a time-out if he or she is able to sit without disrupting others' experiences. If the bad behavior persists, you may be obliged to return to the classroom early. Even so, it is important to give the student a chance to redeem himself or herself the next time you go outdoors.

Routines

Routines are good for management. They impose a measure of self-monitoring because they represent behaviors that are already known and have been practiced. If one person transgresses during a routine, other students are able to intervene to help you with student management. Here are a few routines that may work for you.

Science door. Have you ever watched a group of students pass through the exterior door on their way to recess? As soon as one foot hits the asphalt, they start running and cheering. It is a beautiful sight. Clearly, this is not how you want students to leave the building as you head out for science. One subtle but effective way to distinguish science from recess is to use a different door for science than for recess. Refer to this exit as the "learning door."

Transition behavior. Be explicit about how you want students to walk through the hallways and into the schoolyard. If students exit wildly, simply ring the bell, have them line up inside, and try it again. If this continues to be a problem, return to the classroom and try another day. By consistently showing students that this behavior limits their time outdoors, they will follow your directions.

Taking FOSS Outdoors

Taking FOSS Outdoors

Home base. Establish a destination in the schoolyard where every outdoor activity will begin. Students should walk directly there after leaving the school building. Choose a place that is level and, if possible, away from classroom windows and popular recess areas.

Sharing circle. When students arrive at home base, they should form a large sharing circle—everyone in a single ring with no students hanging back. This is an effective way to maintain eye contact with all students while you give instructions or share findings. Take a position in the circle where you are facing the Sun. This way, you will know that students won't be distracted by having the Sun's glare in their eyes. A sharing circle is also used to transition from one task to another, to summarize an activity, or anytime you need to regroup.

Techniques for forming a circle vary. One method is "magnetic feet." Students spread their legs to meet their neighbors' feet. Magically, these magnets turn off when you direct them to do so. Students may also stand with hands on hips, elbows touching with neighbors'. Pick or create a method that works with your students.

To speed up the formation of a sharing circle, try the tried-and-true countdown from five, with the objective that everyone is in a proper circle by zero.

Attention signal. Adopt a uniform signal for attention. It is essential that students respond to the attention signal immediately. You may choose the same method you use in the classroom or, if this is not appropriate for the outdoors, try one of these.

- A chime, bell, whistle, or other singular and loud sound. These are appropriate outdoors. When students hear it, they stop, look, and listen.

- Count down from five and when you get to one, students are silent with their hands up. This might not be appropriate outdoors. A countdown from ten can be used to call students back to a sharing circle.

- Clap call and response. You clap a pattern, and students return it by repeating the clapping pattern. This works if students are all nearby.

Focus question. Inquiry-based activities are guided by a question. This pedagogical routine should extend into the schoolyard, too. *Students need to know why they are engaged in the outdoor investigation.* They should expect to write the focus question in their notebooks at the outset of the investigation and produce an answer at the end of the investigation.

Boundaries. Setting boundaries allows students freedom within a defined space. Because different activities may require different locations, it is always important to be explicit about where students are allowed to travel during the outdoor activity.

Buddy system. You may want to institute a buddy system, particularly if you leave the schoolyard. When participants are paired off, tell them that each individual is responsible at all times for the whereabouts and safety of his or her buddy. It is helpful and fun to number the pairs in order to count off quickly and account for everyone.

Considerations for Students with Disabilities

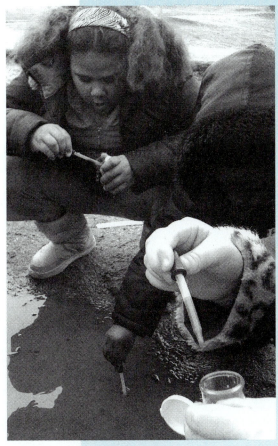

FOSS evolved from pioneering work done in the 1970s with students with physical disabilities. The legacy of that work is that FOSS investigations incorporate multisensory methods, not only to accommodate students with physical and learning disabilities, but also to maximize information gathering for all students. Strategies that provide opportunities to learn for students with disabilities turn out to be good strategies for all students.

All students benefit from opportunities to experience the natural world outdoors. For students with disabilities, consider how to make the schoolyard accessible and safe so that they can work with a degree of independence. This requires advance planning to make sure that the student, his or her family, the special education teacher, and others involved in the child's school experience are informed and have input into the process.

Whenever a student with a disability is successful in a full-inclusion classroom, there is a behind-the-scenes collaborative effort of caring educators who work together to support the student with just the right amount of scaffolding. In advance of teaching the first outdoor activity, contact the special education teachers in charge of each student's Individualized Education Plan (IEP), and have them review the planned outdoor experiences. Ask the teachers to recommend modifications that will better accommodate and support each student. Invite the special educators to join the class for the outdoor activity.

Taking FOSS Outdoors

Taking FOSS Outdoors

Attention and language-based disabilities. Students with attention and behavioral issues often thrive when they are engaged in science outdoors. A fenced area will help you to both keep track of students and provide a sense of safety. Having students work with partners (buddies) allows students to look after each other. Provide short, structured opportunities for students to participate in outdoor activities in a clearly defined space, and expand the boundaries and time expectations as students earn your trust and confidence. Many educators have found great success with treating the outdoor activity as a reward for excellent behavior indoors.

Consider students' communication requirements, and plan to bring specialty devices outdoors with you. This might be as simple as some picture cue cards to help enhance your message or an electronic communication device such as a computer.

Physical disabilities and visual impairments. In the Getting Ready section of each outdoor activity, we ask that the teacher decide where the outdoor activity should be taught. You may find that certain locations are better than others for the purposes of developing science concepts and meeting the physical needs of students with disabilities. Get to know your schoolyard really well, and try to experience it as your students do.

One side of most schoolyards is typically a parking lot, and the other three sides have spaces accessible to students for work and play. If you have a student with a physical disability, you need to consider if the terrain provides for good mobility for the student. Often, schoolyards are handicapped accessible because they are covered in asphalt. For many of the FOSS outdoor activities, an asphalt area is appropriate to use. When you want to use a greener location, make sure wheelchairs or crutches will work on these new surfaces. If the surface will be a mobility challenge, see if a paraprofessional or an educational assistant is able to help the student. If someone is not able to join you, consider if a classmate can help. If this is not an option, then consider working at a transition zone where the grass meets the asphalt.

A student with a visual impairment should make a scouting trip to the outdoor site with a mobility instructor to get the lay of the land and to learn where things are located. If the student becomes familiar with and knows how to navigate his or her outdoor surroundings, it will allow for more independence. Even so, during the actual outdoor activity, the student may need someone to quietly describe the terrain ahead and may need a fellow student's arm for balance and security.

If a student struggles with gross motor coordination, uneven ground may present a challenge. Just as you would in the classroom, begin by offering more support, and slowly pull back on this assistance as students become more comfortable with their stamina, security, and endurance with regular outdoor activities.

Sensory sensitivity. For a student with tactile-sensitivity issues, make it clear that he or she may, for example, observe as a classmate digs in the soil to collect a sample. Over time, this student may feel better able to participate by using gloves or by washing his or her hands as soon as the digging is complete. Knowing where each student falls on the continuum of a disability will help you decide when to hold back asking a student to fully participate, when to allow him or her to just observe, and when to give a gentle nudge and expect more active participation.

For students with sensory disorders, the outdoors is often a calming space. Consider where the quietest place in the schoolyard is, and use this more often if you have students with sensitivity to noise.

No matter what the disability, educators have found success taking students outdoors. With advance planning, communication with the student and the special education team, and a little extra effort, you, too, can provide a rich, safe outdoor learning experience for all your students.

Taking FOSS Outdoors

Taking FOSS Outdoors

TEACHING STRATEGIES

In the beginning, you may find that students regularly use descriptive terms such as "icky," "yucky," and "gross." You may have students who say things such as "I cannot get my clothes dirty. My mom will be mad." Many students are fearful of bugs, wooded areas, and even just sitting on the grass. Often, after a few outdoor activities, these fears and excuses fade away. With patience, persistence, and support, students' resistance may be overcome entirely. If you suspect that your students may be reluctant to work outdoors, structure your first few activities to be low-stress activities. The first few times outdoors can be fairly benign activities with students choosing a comfortable place to just sit (or stand), practicing writing outdoors, and doing simple collecting or counting tasks.

Set the tone. Many teaching strategies that are effective in the classroom work outdoors, too. For example, at the sharing circle, instead of instinctively talking louder (because it is noisy outdoors), kneel down and speak in a loud whisper so that students need to focus to hear you. If students are speaking, put up your silent signal, and wait for silence. The educator's voice sets the tone for the activity.

Take a position. In the sharing circle, position yourself where you have the Sun in your face so that students don't need to squint. If possible, place yourself next to those students who might benefit from a silent look or hand on the shoulder to remind them to be silent.

Meet the challenge. Students who struggle with behavior problems often respond well outdoors when given responsibility. Let the active student carry the heavy jug of water or take the position at the front of the line to lead the class outdoors. For many students, this is all it takes to get them off on the right foot for the outdoor activity.

Students who have the greatest difficulty controlling their behavior indoors are often the leaders when it comes to working in an outdoor space. You may find that students who are not as attentive or cannot sit still inside are the most insistent about quieting down so that the class can get outdoors for science.

Get them writing. Primary students (grades K–2) can fill out a chart on a clipboard outdoors. They are also capable of recording observations outdoors in their notebooks if observation is their only task. Most primary students will need to sit down with their clipboards on their laps or on the ground to do this successfully. In the early years, most writing follows an outdoor activity and is done inside on desks and with the classroom's word wall.

Upper-elementary students (grades 3–5) are capable of writing outdoors. Students will benefit from a quick activity about how to place the notebook or clipboard in the crook of their nonwriting arm for support.

Depending on the activity, you might decide to have students attach their notebooks to a clipboard and place the clipboards in a crate for easy transport and storage. This technique is useful when the ground is moist, when the activity is messy, or when students need to use their hands to complete the activity. The recording will happen immediately after the hands-on activity. Be open to the surprise of how much your students are capable of noticing and recording during and after an outdoor activity.

Taking FOSS Outdoors

FLOW OF OUTDOOR ACTIVITIES

The natural flow of a FOSS outdoor activity is slightly different from that of a standard FOSS indoor activity. The steps of a typical outdoor activity are listed below. This list may be helpful if you want to teach more than the handful of outdoor activities in the *Investigations Guide*, or if you want to adapt an indoor activity for schoolyard use.

1. **Prepare for the outdoor activity.**
   - Determine the best location to teach the activity.
   - Check the weather forecast.
   - Make sure students will be dressed appropriately.
   - Prepare materials for distribution.
   - Check the site the morning of the activity.

2. **Set the learning objective.**
   - Present the focus question.
   - Discuss procedures.

3. **Go outdoors.**
   - Gather at the predetermined location.

4. **Model or describe the activity.**
   - Organize students.
   - Define boundaries.
   - Introduce/distribute materials.

5. **Monitor the activity.**
   - Check student engagement.
   - Check student recording.
   - Ask questions.

6. **Share the experience.**
   - Form a sharing circle to discuss experiences.
   - Share thinking.
   - Share answers to the focus question.

7. **Return to class**
   - Make connections to the related indoor activity.
   - Display student work and collections.

EXTENDING BEYOND FOSS OUTDOOR ACTIVITIES

Occasionally, you may stumble upon a serendipitous opportunity. A breeze may launch thousands of twirling seeds from a maple tree, a woodpecker may alight on a tree so close that students can observe it drumming for insects, student-made parachutes may be carried by an updraft high into the sky and out of sight. To your delight, you may spy something you have never seen before. It can happen at any time when you are outdoors!

At special moments like these, our job as educators is to signal students to stop and quietly appreciate the suspension of time. Sometimes, words break the wonder. Trust your instincts at magical moments like these. The answers to questions will come eventually. It is not essential to label the event or even understand it. By inviting students to be alive with their feelings in the moment, you give them a gift for a lifetime.

It is not uncommon for educators to experience the powerful effect of the outdoors on student learning. If you find yourself searching for other outdoor learning opportunities, consider the ones below.

Move activities outdoors. Whatever the subject, students will have more room outdoors to be creative with some activities, and you can worry less about water, sand, and gravel spills. You must still consider how to transport materials, where students will sit, how they will return their project to the classroom, and how to clean up the outdoor space and students' hands before returning to the building.

Use the outdoors for extensions. Extending an inside concept to the outdoors is an excellent way to apply new knowledge. For example, in the **Structures of Life Module**, students grow bush beans hydroponically. If the large leaves fascinate students, go outdoors and see how many kinds of leaves you can find in the schoolyard. Do they all have smooth edges and come to a point at their tips? Go on a leaf hunt, group the leaves by their characteristics, and, eventually, have students tape them into their science notebooks.

There is great value in repeating an indoor activity outdoors. If your students are sanding wood samples inside, follow this up with a trip outdoors to find a stick and sand it. Have you been studying sow bugs? Ask students if they think they know where in the schoolyard they might find these bugs. Applying what students have learned in the classroom and putting that knowledge to work outdoors is an effective way to solidify their understanding. It's also an effective way to informally assess whether students understand the concepts, as well as a method for reinforcing the learning.

Taking FOSS Outdoors

Taking FOSS Outdoors

> **NOTE**
> Remember not to release any classroom animals into schoolyard environments.

Find solitude. Use your outdoor space for silent independent work time. Just as in the classroom, the outdoor space can be a workspace with activities going on. At times, the outdoor space is more of a sanctuary for independent observation and notebook writing. It can be a place for special classroom rituals, awakening awareness of the beauty of nature. Sometimes, it can just be a place to be silent for a minute to awaken the senses and refocus students' energy. Some teachers increase this silent minute to 2, 3, or even 5 minutes. Silence is something to be practiced, and for many students and teachers, this can be challenging. This is a special way to end an outdoor experience and will help students transition into the classroom.

Enhance biodiversity. Modify your schoolyard by adding natural materials, such as logs, rocks, or paving stones. These structures can provide safe havens that may attract more living things. These types of shelters can be particularly helpful if you have an environment without natural shelter from the Sun, such as trees and shrubs. Students can also be involved in the design and implementation of these projects.

Schoolyard modification of this kind requires administrative participation and the support of the school custodian. Marking the area with educational signage can further benefit the enhanced site. If your schoolyard habitat needs your intervention to cultivate biodiversity, understand that it can take a couple of years to get established. Areas completely surrounded by blacktop or concrete can become filled with living things if provided with food, shelter, and water.

Attract wildlife. There are many responsible ways to attract wildlife to your class windows with feeders for birds, squirrels, hummingbirds, or butterflies, as well as many great programs for monitoring these animals. See FOSSweb for ideas for additional wildlife observation projects.

Establish long-term studies. The possibilities for long-term studies are endless, ranging from weather monitoring to seasonal population variation. It can be as simple as adopting an observation location and visiting it monthly to monitor various aspects of change over time. See FOSSweb for ideas for long-term projects.

Create gardens. Planting a garden in raised beds or improved soil is an ambitious option for increasing the biodiversity of your schoolyard. Consider carefully, especially with a vegetable garden, the timing of the school year. In most parts of the country, the time when plants require the most support is during summer vacation. Even if you can get a summer program involved, we suggest starting with indigenous plants that bloom or mature in spring and fall and require little maintenance.

ELEMENTARY-LEVEL ENVIRONMENTAL EDUCATION

In the early 1990s, David Sobel noticed something poignant about children's perceptions of the environment. If a child had been introduced to environmental issues at school that were presented in the context of doom-and-gloom scenarios, the child expressed a heightened sense of anxiety and hopelessness, which Sobel calls ecophobia (Sobel 1996). The implications of his finding should raise a cautionary flag. Sobel is not suggesting that we abandon teaching about the environment in our elementary schools. He is proposing a different approach to environmental education that will bring our children into natural, healthy relationships with environmental issues.

Effective early environmental education should focus on local and ultralocal issues. What is happening in our schoolyards? What factors influence the communities of plants and animals in our neighborhoods? How do changing weather conditions affect the populations around our schools? How are our actions affecting the habitats in our schoolyards? What can we do to enhance natural systems at our schools? Elaborate rain forest projects provide little understanding and have negligible impact on students' connections to nature; researching and installing a butterfly garden or keeping an inventory of the birds in the schoolyard can be transformative. The children from Sobel's 1996 study could tell you how many species in the Amazon were going extinct each minute, but were unfamiliar with the most common plants in their schoolyard.

Time outdoors during the school day is beneficial for student learning. Students who are exposed to hands-on experiences in their local environment often become enthusiastic, self-motivated learners and, typically, academically outperform their peers who do not have these learning opportunities (Liebermann and Hoody 1998). Children are able to pay attention for longer periods of time outdoors on the same assignment and are more focused when they return to their indoor class work (Louv 2008).

Research has produced evidence that using the schoolyard is an effective way to enhance student learning. Texas A&M University, in conjunction with the Texas Education Agency, conducted a meta-analysis of the research in order to identify and rank effective instructional methods for science education and to define how best to improve student achievement. The highest-ranked teaching strategy was Enhanced Context Strategies, which included taking meaningful field trips and using the schoolyard for activities (Scott et al. 2005).

Taking FOSS Outdoors

Taking FOSS Outdoors

Students' attitudes toward learning are influenced by simple outdoor experiences. In one study (Shaw and Terrance 1981), students who experienced outdoor instruction reported that, in general, they enjoyed school more and felt more supported and trusted by their teacher than they had prior to the outdoor experiences. These pretest/posttest differences were more pronounced for students who had been identified as being "uninvolved" in the classroom activities. Also, this student perception was a lasting effect that carried over to the regular classroom activities weeks later.

Perhaps the most important benefit of incorporating the outdoors into the traditional school learning environment is that it offers opportunities for students to synthesize concepts and personal experience by applying what they have learned to a new environment.

FOSS outdoor activities will help you focus on age-appropriate environmental topics and enable you to create meaningful and personal connections between your students and their local environment. When students can openly explore the environment, they can create meaningful connections to their learning and establish positive relationships with nature. You'll be amazed by what students notice.

Here's the good news. If you focus on inquiry and direct experience instead of problems, it takes remarkably little guidance for students to make positive, empowering, lifelong connections to nature. One insightful young man explained, "My video games have a pattern that is always the same, but nature is like a game that is different every time you play." As an educator, you can draw out that sense of wonder and curiosity for students while simultaneously helping them build a solid science foundation.

REFERENCES

Liebermann, G., and L. Hoody. 1998. *Closing the Achievement Gap: Using the Environment as an Integrated Context for Learning; Results of a National Study.* San Diego: State Education and Environment Roundtable.

Louv, R. 2008. *Last Child in the Woods: Saving Our Children from Nature-Deficit Disorder.* New York: Workman Publishing.

Scott, T. et al. 2005. *Texas Science Initiative Meta-Analysis of National Research Regarding Science Teaching.* Texas Education Agency.

Shaw, T. J., and J. M. Terrance. 1981. "Involved and Uninvolved Student Perceptions in Indoor and Outdoor School Settings." *Journal of Early Adolescence,* 1:135–146.

Sobel, D. 1996. *Beyond Ecophobia: Reclaiming the Heart in Nature Education.* Great Barrington, MA: Orion Society.

Taking FOSS Outdoors

ACKNOWLEDGMENTS

The Taking FOSS Outdoors initiative got its start through a collaboration with the Boston Schoolyard Initiative (BSI). In 2004, BSI began developing an approach to teaching science that routinely takes students into the schoolyard to test, apply, and explore core science concepts and skills. As part of this project, BSI developed *Science in the Schoolyard Guide*s™ for 12 FOSS modules and a companion *Science in the Schoolyard*™ DVD. In partnership with the City of Boston, BSI designs and builds schoolyards that provide a rich environment for teaching, learning, and play. For more information on BSI, *Science in the Schoolyard*, or BSI's *Outdoor Writer's Workshop*™ professional development program and materials, see www.schoolyards.org.

Science Notebook Masters

"Physical Systems" Video Review

1. What effect did the eruption of Mount St. Helens have on the geosphere, atmosphere, hydrosphere, and biosphere of the region?

2. What is an ecosystem?

3. In what ways do people affect the balance of production and consumption within an ecosystem?

4. What was the dust bowl?

5. What are invasive species? Why are they considered one of the greatest threats to an ecosystem?

6. When is a system said to be in a state of equilibrium?

7. What are renewable resources? Provide some examples.

Woods Ecosystem

Make several food chains of at least three organisms. Use arrows to show how the energy of food moves from organism to organism.

- American robin
- Aquatic snail
- Bacteria
- Black bear
- Brook trout
- Chipmunk
- Coyote
- Dead plants and animals
- Earthworm
- Grama grass
- Great blue heron
- Green algae
- Grouse
- Hare
- Mayfly
- Pine trees
- Red-tailed hawk
- Scuds
- *Tubifex* worm
- Wild blueberry

Kelp Forest Food Web

Make a food web using all the organisms in the kelp forest ecosystem.

- Bat star
- Garibaldi
- Giant kelp
- Kelp crab
- Marine bacteria
- Phytoplankton
- Purple sea urchin
- Red octopus
- Sea otter
- Señorita fish
- Turban snail
- Zooplankton

Kelp Forest Food Web

Make a food web using all the organisms in the kelp forest ecosystem.

- Bat star
- Garibaldi
- Giant kelp
- Kelp crab
- Marine bacteria
- Phytoplankton
- Purple sea urchin
- Red octopus
- Sea otter
- Señorita fish
- Turban snail
- Zooplankton

Response Sheet—Investigation 1

A student drew this food web in his notebook. Another student was looking at it and said, "I agree with the organisms you've used for the food web, but I disagree with the direction you drew arrows. I also think you are missing something. Food webs usually include producers, consumers, and the Sun."

If you were a third student taking part in this conversation, what would you tell the other two?

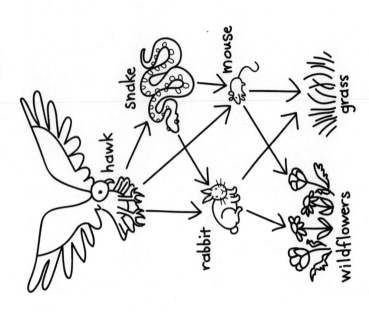

Making a Redworm Habitat

1. Put about 1–2 centimeters (cm) of garden soil in the jar.

2. Tear two large sheets of newspaper into thin strips. Moisten the paper strips with water. They should be moist, but not dripping wet. A spray mister is a good way to moisten the newspaper.

3. Fill the jar with the damp newspaper strips until it is almost full, about 6–7 cm from the top.

4. Add some natural leaf litter (five or six dead leaves, two or three dead twigs) and a small amount of fresh household waste (apple cores, lettuce scraps, crushed eggshells, coffee grounds, melon rinds, etc.).

5. When the materials are all in the jar, screw on the lid and give the container a shake to mix the contents a bit.

6. If necessary, use a spray mister to moisten the habitat. The contents should be very moist, but not dripping wet.

7. When the container is ready, count 15–18 redworms and drop them into the container. Screw on the lid. It has air holes for ventilation.

Activating Yeast

What does yeast need to break its dormancy?

1. Get two 1-liter zip bags. Label one bag "cookie."
2. Put two level 5-milliliter (mL) spoons of yeast into each zip bag.
3. Use a syringe to put 50 mL of hot water in each bag.
4. Put two animal crackers in the "cookie" bag, and nothing in the other bag.

Zoo Parade Cookies
Ingredients: Wheat flour, sugar, partially hydrogenated vegetable shortening, whole eggs, butter, high-fructose corn syrup, salt, vanilla, baking soda, whey.

Choco-Chunk Cookies
INGREDIENTS: Wheat flour, sugar, sweet chocolate, corn syrup, partially hydrogenated vegetable shortening, nonfat milk, cornstarch, invert syrup, vanilla, pectin, baking soda, salt, citric acid, caramel color.

Big Bite Snaps
Ingredients: Unbleached wheat flour, sugar, milk chocolate, partially hydrogenated vegetable shortening, whole eggs, brown sugar, nonfat milk, butter, baking soda, egg whites, vanilla, salt.

Package labels list the ingredients in order from most to least by quantity. What two ingredients are present in the greatest quantity in the cookies?

- Zoo Parade Cookies
- Choco-Chunk Cookies
- Big Bite Snaps

Wheat-Seed Investigation

1. Fill four 1/2-liter containers (planters) almost full with soil, using about 1.5 cups of soil.
2. Sprinkle one 5-milliliter (mL) spoon of wheat seeds over the surface of the soil (1 spoon is about 100 seeds).
3. Sprinkle an additional 50 mL of soil to cover the seeds.
4. Pour 100 mL of water carefully over the planted seeds.
5. Close two of the planters in clear plastic bags. Small binder clips can be used to close the top.
6. Close the other two planters in black plastic bags.
7. Place the four bagged planters in a warm, lighted location.

Wheat-Seed Investigation

1. Fill four 1/2-liter containers (planters) almost full with soil, using about 1.5 cups of soil.
2. Sprinkle one 5-milliliter (mL) spoon of wheat seeds over the surface of the soil (1 spoon is about 100 seeds).
3. Sprinkle an additional 50 mL of soil to cover the seeds.
4. Pour 100 mL of water carefully over the planted seeds.
5. Close two of the planters in clear plastic bags. Small binder clips can be used to close the top.
6. Close the other two planters in black plastic bags.
7. Place the four bagged planters in a warm, lighted location.

"Getting Nutrients" Review

1. What is the difference between heterotrophs and autotrophs?

2. What does food provide for organisms?

3. A food pyramid describes levels in a feeding relationship involving producers, consumers, and decomposers. What information does a food pyramid describe that a food web might not?

4. What level of consumer are humans?

"The Human Digestive System" Review

1. What is digestion?

2. How might the human and painted lady butterfly digestive systems be similar?

3. Why do you think the digestive system is called a system?

Experiment on Chemical Digestion in the Stomach

The students in the video conducted an experiment to find out what happens to food (hard-boiled egg white) in different environments. Talk in your groups about this experiment.

1. What was the question?

2. What was controlled, and what changed in the experiment?

3. What were the results?

4. What was the conclusion?

Response Sheet—Investigation 2

A student said, "I have a model of how digestion works! In the stomach, nutrients are made into food. The food is then used by the cells for energy."

Do you agree with this student? If not, what information can you provide to this student to clarify his understanding of digestion?

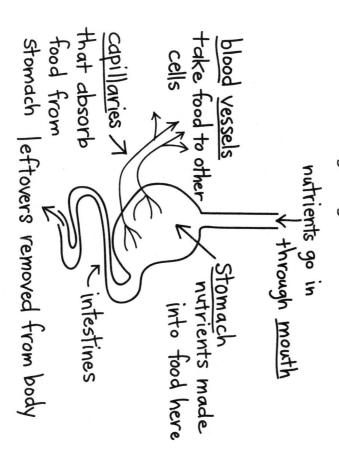

Wheat-Seed Chamber Setup

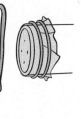
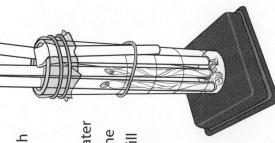

1. Stretch the small piece of plastic film across the top of a graduated cylinder, and secure it there with a rubber band, like a drum head across the top of the graduated cylinder.

2. Use a sharp pencil to carefully poke four holes in the drum head.

3. Carefully insert three wheat straws through three of the holes, with the paper-towel side down, inside the graduated cylinder.

4. Insert a plain clear straw into the fourth hole in the drum head.

5. Use a syringe to carefully introduce water into the graduated cylinder through the plain plastic straw. The water should fill the graduated cylinder only up to the 10 mL mark. After the water is in the graduated cylinder, bend the plain straw over along the side of the graduated cylinder, and secure with a rubber band.

"Plant Vascular Systems" Review

1. What kinds of vessels are in leaf veins?

2. How does water get to the cells at the top of a plant?

3. What is sap? What does it do?

"The Human Circulatory System" Review

1. What is the heart and what is its role in the circulatory system?

2. What are heart valves and what do they do?

3. Where are the heart valves?

4. What is the main function of the left side of the human heart?

5. What is the main function of the right side of the human heart?

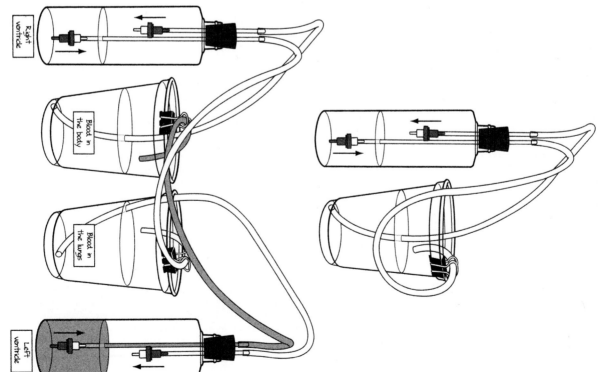

Response Sheet—Investigation 3

Both plants and animals use systems to transport materials to and from their cells. Compare the vascular system of a plant to the circulatory system of a human. How are they alike and how are they different?

Response Sheet—Investigation 3

Both plants and animals use systems to transport materials to and from their cells. Compare the vascular system of a plant to the circulatory system of a human. How are they alike and how are they different?

"The Human Respiratory System" Review

1. What are the parts of the respiratory system? What is the system's function?

2. What are alveoli and what happens there?

Measuring Vital Capacity

1. Measure and record your vital capacity (lung volume) three times.
2. Calculate your average vital capacity.

Trial	Vital capacity (L)
1	
2	
3	
Average	

Measuring Vital Capacity

1. Measure and record your vital capacity (lung volume) three times.
2. Calculate your average vital capacity.

Trial	Vital capacity (L)
1	
2	
3	
Average	

Measuring Vital Capacity

1. Measure and record your vital capacity (lung volume) three times.
2. Calculate your average vital capacity.

Trial	Vital capacity (L)
1	
2	
3	
Average	

Measuring Vital Capacity

1. Measure and record your vital capacity (lung volume) three times.
2. Calculate your average vital capacity.

Trial	Vital capacity (L)
1	
2	
3	
Average	

"Structures of the Brain" Review

1. Describe the components of the central nervous system. How is it a system?
2. What functions does the brain stem control?
3. What are sensory neurons, and what is their role?
4. What are motor neurons, and what is their role?

Stimulus/Response

Height of drop _____
Stimulus _____
Response _____

	Hit	Miss
5		
4		
3		
2		
1		

Stimulus/Response

Height of drop _____
Stimulus _____
Response _____

	Hit	Miss
5		
4		
3		
2		
1		

Stimulus/Response

Height of drop _____
Stimulus _____
Response _____

	Hit	Miss
5		
4		
3		
2		
1		

Stimulus/Response

Height of drop _____
Stimulus _____
Response _____

	Hit	Miss
5		
4		
3		
2		
1		

Attention Action Card

Name _____

I respond to these two colors: _____
and _____.

I respond to this pattern:
☐ Squares ☐ Spots
☐ Diamonds ☐ Rectangles
☐ Triangles ☐ Stripes

I prefer this habitat:
☐ Grass ☐ Bushes and trees ☐ Arid, rocky soil

Attention Action Card

Name _____

I respond to these two colors: _____
and _____.

I respond to this pattern:
☐ Squares ☐ Spots
☐ Diamonds ☐ Rectangles
☐ Triangles ☐ Stripes

I prefer this habitat:
☐ Grass ☐ Bushes and trees ☐ Arid, rocky soil

Attention Action Card

Name _____

I respond to these two colors: _____
and _____.

I respond to this pattern:
☐ Squares ☐ Spots
☐ Diamonds ☐ Rectangles
☐ Triangles ☐ Stripes

I prefer this habitat:
☐ Grass ☐ Bushes and trees ☐ Arid, rocky soil

Attention Action Card

Name _____

I respond to these two colors: _____
and _____.

I respond to this pattern:
☐ Squares ☐ Spots
☐ Diamonds ☐ Rectangles
☐ Triangles ☐ Stripes

I prefer this habitat:
☐ Grass ☐ Bushes and trees ☐ Arid, rocky soil

"Animal Communication" Review

Discuss these questions in your group.

1. Why would an organism want to call attention to itself?
2. What purpose does the rattlesnake's rattle serve?
3. What purpose does sweet scent serve for a plant?
4. What plant or animal has features or behaviors other than color and pattern to attract attention?
5. How is an attention mechanism a stimulus/response interaction?

Write answers to these questions in your notebook.

6. What is communication?
7. What kinds of stimuli can initiate communication?
8. What kinds of things do animals need to communicate?

Response Sheet—Investigation 4

When woodpeckers tap loudly on dead tree trunks, it is called drumming. The sound can carry a great distance in the forest.

Why might a woodpecker drum? How would you explain this behavior in terms of a stimulus/response adaptation?

"Animal Behavior and Communication" Video Review

1. How do dogs learn to wag their tails when they are "happy"? Who teaches kittens to chase strings and other small toys?

2. What instinctive behaviors do sea turtles engage in that help them survive?

3. How do bees, ants, and termites learn how to build their complex communities?

4. What are three instinctive behaviors that some animals exhibit to deal with harsh environmental conditions?

"Monarch Migration" Review

1. Think about the monarch migration system. What are the parts?

2. What might happen if the life cycle of the milkweed plant changed in some way?

"North Atlantic Ocean Ecosystem" Review

1. What are the major producers in the North Atlantic Ocean ecosystem?

2. What is a phytoplankton bloom?

3. Why is the North Atlantic bloom important to study?

4. Describe some of the instrumentation scientists use to study the North Atlantic bloom.

Teacher Masters

LETTER TO FAMILY

Cut here and paste onto school letterhead before making copies.

―――――――――――――――――――――――――――――――――――――

═════════════════ **Science News** ═════════════════

Dear Family,

Our class is beginning a new science unit using the **FOSS Living Systems Module**. We will be looking at a number of life science concepts from the standpoint of a system—a collection of interacting parts. We will start with ecosystems and describe the feeding relationships in food chains and food webs involving producers (plants and algae), consumers, and decomposers. To investigate the role of decomposers, we will set up redworm habitats and provide organic materials in the form of shredded moist newspaper with a small amount of selected fresh kitchen scraps and leaf litter integrated into the habitat. We will observe the action of the redworms for 8 weeks and analyze the result of their activity at the end of the module.

We will also investigate transport systems in multicellular organisms that provide each cell with food, water, gas exchange, and waste removal. Students will learn about the structures, functions, and interactions of the digestive, circulatory, and respiratory systems in humans. We will build model heart systems and investigate vital capacity of our lungs. Students will learn about the vascular system in plants (xylem and phloem), and they will compare that system for moving water, minerals, and sugar to the transport system in humans. They will also be introduced to the process of photosynthesis. Students will be designing and conducting controlled experiments to investigate some of these systems: use of sugar by yeast cells and light by sprouting wheat seeds.

We will also learn about the central nervous system in humans and compare it to other animals. With the focus on systems, we will investigate stimulus/response in human response time and in animal communication and behavior. We will also be looking at instinctive behavior, such as that exhibited by migrating monarch butterflies.

Watch for the home/school connection sheets I will be sending home with your child. These suggest ways for the whole family to investigate interesting aspects of our life science study. In addition, you and your child can visit the FOSS website (www.fossweb.com), where there are instructional activities, interactive simulations, and resources related to the Living Systems Module.

If you have any questions or comments, please drop me a note or come in and visit our class. We are looking forward to many weeks of exciting investigations.

Sincerely,

Name _____ Date _____

MATH EXTENSION—PROBLEM OF THE WEEK
Investigation 1: Systems

1. A class has decided to use a worm bin to vermicompost lunch scraps. One student is in charge of figuring out the size of the worm bin. She found out that they need a surface area of 900 square centimeters (cm) for every 500 grams (g) of food scraps per week. The class produces about 1.5 kilograms (kg) of food scraps a week.

   Design a worm bin that would have a large enough surface area to vermicompost the class lunch scraps. Show your work, including a drawing that shows the length and width of the bin. (The height of the bin can be 20 cm.)

2. Another student is in charge of figuring out how many redworms the class needs to put in the worm bin. This is what he knows about the food habits of redworms and his classmates.
   - Redworms can eat about half of their weight in food every day.
   - One thousand (1,000) worms weigh about 0.5 kg.
   - Each student produces about 25 g of lunch scraps a day.
   - There are 30 students in the class.

   What is the least number of worms they need for the worm bin? Show your work.

3. The class worm bin is 3 months old. There are so many more worms now! One science group wants to know how fast redworms reproduce. This is what they know about the redworm life cycle.
   - Redworms have both eggs and sperm and can produce cocoons.
   - Redworms can produce cocoons at 3 months old.
   - Redworms produce 3 cocoons per week.
   - It takes 11 weeks for the cocoons to hatch.
   - Each cocoon produces an average of 3 baby redworms.

   The group puts 2 worms in a worm bin and observes them for 3 months.

   How many cocoons might the group find in the worm bin after 11 weeks? How many baby worms? Show your work.

   How many cocoons would there be after 12 weeks? How many baby worms? Show your work.

   If the class worm bin started with 1,000 worms, how many baby worms might they have after 3 months?

HOME/SCHOOL CONNECTION
Investigation 1: Systems

The redworm or red wiggler (*Eisenia fetida*) is a very thin and relatively small earthworm, about 3–8 centimeters (cm) long. This is the common species of redworm used in vermicomposting. Although the redworms are small, they can shred and consume nearly half their weight in food every day. They eat decaying leaves and other decaying plant parts that have been broken down by the action of bacteria, fungi, and microorganisms. Actually, their main food source is the bacteria, fungi, and other microorganisms on the decaying plant matter, but they also consume the organic matter on which the microorganisms live. In worm bins, redworms feed on grass clippings and kitchen scraps, including vegetables, fruit, egg shells, coffee grounds, paper, and cardboard.

Here are some of the kinds of kitchen waste that redworms eat: potato peels; coffee grounds with filters; tea bags; apple cores; crushed egg shells; stale bread; parings of cucumber, carrots, squash, lettuce, melon rinds, and so forth. If potatoes are used, they should be cooked.

Here are some of the things that are not good for redworms: meat, fat, oil.

Make a list of the kitchen waste that your home produces that redworms could eat and would be good to use in a worm bin. Keep the list for at least a week.

Name _____ Date _____

MATH EXTENSION—PROBLEM OF THE WEEK

Investigation 2: Nutrient Systems

You get energy from food.

Three classes of nutrients provide energy: carbohydrate, protein, and fat.

Food energy is measured in calories. You get calories from carbohydrate, protein, and fat.

You get different numbers of calories from different nutrients.

    1 gram of carbohydrate = 4 calories (Cal)

    1 gram of protein   = 4 calories (Cal)

    1 gram of fat   = 9 calories (Cal)

If you know how much carbohydrate, protein, and fat is in a piece of food, you can calculate how many calories it has. For instance, if a baked potato has 50 g of carbohydrate, 1 g of protein, and 1 g of fat, you can calculate the total calories.

    50 g carbohydrate × 4 Cal/g = 200 Cal

        1 g protein × 4 Cal/g = 4 Cal

            1 g fat × 9 Cal/g = 9 Cal

                      Total = 213 Cal

Problem

A boy went to a baseball game. During the game, he ate a hot dog, a bag of chips, and a soft drink. When he got home, he wondered how many calories he got from his fast-food meal. He looked up the average calories for the items he ate. The data are shown in the table.

Food item	Protein (g) (4 Cal/g)	Carbohydrate (g) (4 Cal/g)	Fat (g) (9 Cal/g)
Hot dog	8	20	16
Potato chips	4	31	20
Soft drink	0	36	0

How many calories was the boy's meal?

The boy was happy with the total calories in his meal. But he wants to have only 30 percent of his calories from fat. Does fat provide more than 30 percent of the calories of his meal?

If the boy has too much fat in his meal, how many grams of fat will he have to remove? How many grams of carbohydrate and/or protein will he have to add?

HOME/SCHOOL CONNECTION

Investigation 2: Nutrient Systems

Listen to your body's internal systems—they make sounds. Use a stethoscope to listen if you have one. You can make a simple listening device with two small plastic cups and a short piece of plastic tubing.

Ask an adult to use a nail to make a small hole in the bottom of both cups. Force the tubing into the holes. It should fit very tightly.

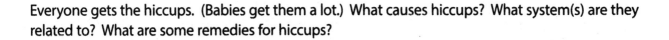

Put the mouth of one cup over the thing you want to hear. Put the other cup over your ear. Listen to your heart, lungs, stomach, intestines, your throat swallowing, and your teeth chewing.

Research these questions, by using reliable sources.

Everyone gets the hiccups. (Babies get them a lot.) What causes hiccups? What system(s) are they related to? What are some remedies for hiccups?

You have felt and heard your own stomach growl. When does it growl? What makes it growl? What system is involved?

Everyone burps now and then. (Babies burp a lot.) What is going on when you burp? What system is involved?

What is a sneeze? What system is involved?

HUMAN HEART VALVES

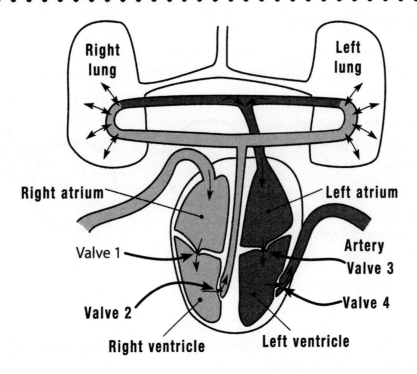

The human heart is a fluid pump. The heart is a system of many parts, including blood, blood vessels, heart chambers, muscles, and one-way valves. The muscles put pressure on the blood inside the chambers. The pressure pushes blood from chamber to chamber and through the blood vessels. The valves let blood flow in only one direction. Valves keep the blood flowing in the right direction.

Valve 1, the tricuspid valve, lets blood flow into the right ventricle as it comes back to the heart from the body.

Valve 2, the pulmonary valve, lets blood flow from the right ventricle as it is pumped to the lungs.

Valve 3, the mitral valve, lets blood flow from the lungs into the left ventricle.

Valve 4, the aortic valve, lets blood flow out of the left ventricle when it is pumped to all parts of the body.

CIRCULATORY SYSTEM MODEL

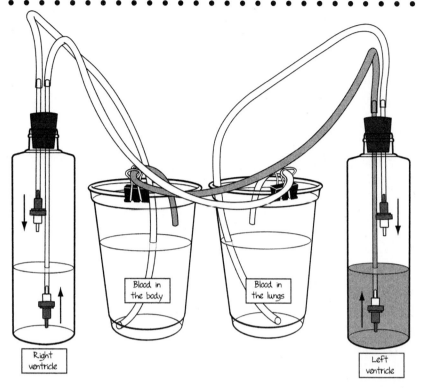

1. Slide one long rigid pipe and one short rigid pipe through the two holes in a rubber stopper. Leave about 2 centimeters (cm) of both rigid pipes sticking up from the stopper.

2. Stick a valve in the end of each pipe that will be inside the bottle. Insert the blue end into the short pipe, and the clear end into the long pipe.

3. Attach a long flexible hose (tubing) to the top of each pipe.

4. Repeat steps 1–3 to make a second ventricle.

5. Position the right ventricle bottle near the large plastic "body" cup. Run the end of the hose connected to the short pipe to the bottom of the "body" cup.

6. Run the end of the hose connected to the long pipe to the top of the "lung" cup. Use a binder clip to keep it in place at the top of the open "lung" container.

7. Arrange the left ventricle hoses in a similar manner: hose from the short pipe to the bottom of the "lung" cup and hose from the long pipe to the rim of the "body" cup.

8. Half fill all containers with water. Squeeze the bottles gently but firmly to operate the model.

LUNG VOLUME BAG CALIBRATION A

Each pair of students will need one lung volume bag for use in measuring their vital capacity. The lung volume bag needs to be labeled in liters (calibrated) in order to accurately measure the amount of air.

Have students calibrate the bags before the activity. The best way to do this is to set up a number of stations in the room and have pairs of students go to a station to prepare their bag. Set up at least two stations, but more if you have the space.

For each calibration station, you will need:

1 Lung Volume Bag Measurements in Liters, prepared from Teacher Master 10
1 Permanent marking pen
• Transparent tape
• Masking tape
1 Meter tape
• A 6 ft side of a table top

Cut out the four measurement pieces from teacher master 10, Lung Volume Bag Measurements in liters, on the dashed lines.

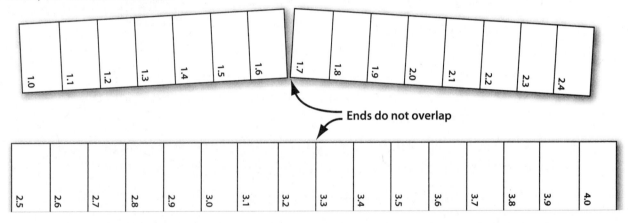

Ends do not overlap

Align the pieces as illustrated, placing the sections in numerical order. Tape the junctions, front and back so you have one long measurement strip.

Label a piece of masking tape "start" and place it on one side of a long table. Tape the Lung Volume Bag Measurements in Liters securely to the table so the edge begins 32 cm from the front of the masking tape.

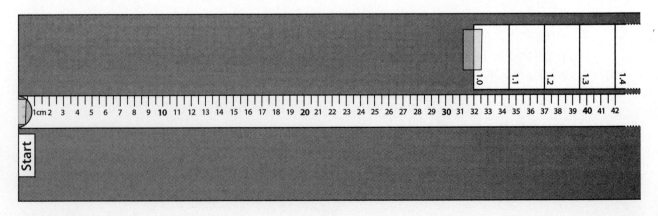

FOSS Next Generation
© The Regents of the University of California
Can be duplicated for classroom or workshop use.

Living Systems Module
Investigation 3: Transport Systems
No. 8—Teacher Master

LUNG VOLUME BAG CALIBRATION B

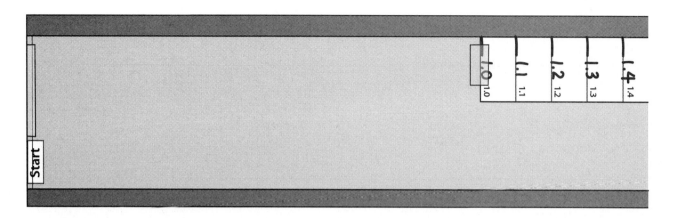

Tape the bottom edge of the lung volume bag to the table to align with the front edge of the masking tape. Pull the lung bag tight to remove any slack and tape the top edge to the table as well. The Lung Volume Bag Measurements in Liters strip should be visible under the bag.

Using a marking pen, label the bag with the correct measurements in liters. Be sure the bag remains flat on the table.

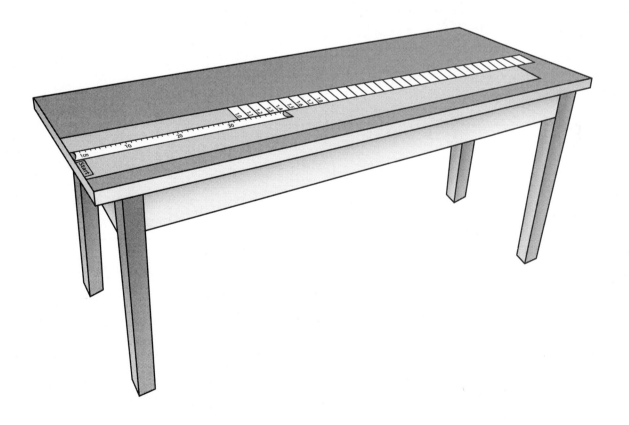

LUNG VOLUME BAG MEASUREMENTS IN LITERS

	2.4	3.2	4.0
1.6	2.3	3.1	3.9
1.5	2.2	3.0	3.8
1.4	2.1	2.9	3.7
1.3	2.0	2.8	3.6
1.2	1.9	2.7	3.5
1.1	1.8	2.6	3.4
1.0	1.7	2.5	3.3

MATH EXTENSION—PROBLEM OF THE WEEK

Investigation 3: Transport Systems

A student (Student A) put a stalk of celery with six leaves in 100 milliliters (mL) of water. The celery leaves were all about the same size. One day later, there was only 70 mL of water left in the cup.

Student A setup at starting time

Student A setup 1 day later

Another student (Student B) put a stalk of celery in 100 mL of water. One day later, only 50 mL of water was left in the cup.

Student B setup at starting time

Student B setup 1 day later

How many leaves were on Student B's celery stalk?

HOME/SCHOOL CONNECTION

Investigation 3: Transport Systems

Celery stalks have vascular bundles. The xylem tubes transport water from the roots (base of the stem) to the leaves. This is how the cells in the celery leaves get water and minerals to stay alive.

Do other vegetables transport water? You can use colored water to find out.

Visit the produce section when you are at the market. Get a few things to test. Try different kinds of cabbage and lettuce, green onions and leeks, asparagus, and other interesting things. Bring the results of your investigations to school to share.

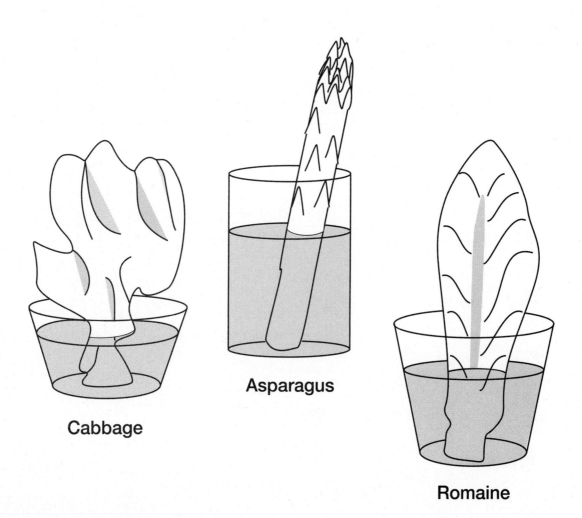

Name _____ Date _____

MATH EXTENSION—PROBLEM OF THE WEEK
Investigation 4: Sensory Systems

Students in a class were testing their arm/shoulder muscle strength by doing chair push-ups. Each student did as many push-ups as he or she could without resting. Here are the results.

Boy A 11	Boy C 15	Boy F 17	Girl H 14
Girl A 16	Boy D 15	Boy G 14	Boy I 16
Girl B 15	Girl D 12	Boy H 14	Girl I 10
Boy B 19	Boy E 20	Girl F 16	Girl J 17
Girl C 14	Girl E 18	Girl G 15	Boy J 13

Graph the results.

What was the total number of push-ups done by the class? _____

What questions do you have that can be answered by analyzing these data?

HOME/SCHOOL CONNECTION A

Investigation 4: Sensory Systems

It is possible to find out how quickly you can respond to a visual stimulus by using a response timer. Tape a pencil to the back of the response timer (paper/strips), as shown in the illustration. The eraser end of the pencil should be flush with the starting-position end of the strip.

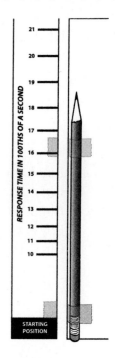

To use the timer, you need two people.

1. One person holds up the timer by the top of the paper strip.

2. A second person, the catcher, positions her or his fingers over the words "starting position," ready to catch the response time the instant it begins to fall.

3. When the catcher sees the strip start to fall, he or she catches it and notes the number under his or her thumb. The number represents the number of 100ths of a second it took to respond.

4. Record on the record sheet your response times for five trials with both your left and right hands. Average the results to get your average response time.

5. Compare the response times for your left and right hands. Explain why you think one hand responds faster than the other.

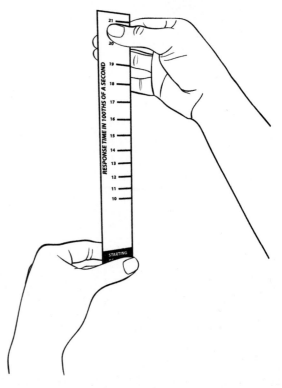

HOME/SCHOOL CONNECTION B
Investigation 4: Sensory Systems

Response time in 100ths of a second: 10, 11, 12, 13, 14, 15, 16, 17, 18, 19, 20, 21 (four identical bar-chart templates)

Starting position | Starting position | Starting position | Starting position

HOME/SCHOOL CONNECTION C

Investigation 4: Sensory Systems

Find out how fast your hand can respond. Start with a visual stimulus. Test your left and right hands five times. Record your response time after each drop.

Stimulus _____	
Response _____ hand	
Drop	Time
1	
2	
3	
4	
5	
Total	

Average _____

Stimulus _____	
Response _____ hand	
Drop	Time
1	
2	
3	
4	
5	
Total	

Average _____

Calculate the average response time for each hand. Write the averages on the lines under the totals.

Which hand had the faster response time? _____

Explain why you think that hand responded faster.

_____
_____
_____

Assessment Masters

Embedded Assessment Notes

Living Systems

| Investigation ___, Part ___ | Date _____ |

Got it!

Concept

Concept

Reflections/Next Steps

| Investigation ___, Part ___ | Date _____ |

Got it!

Concept

Concept

Reflections/Next Steps

Performance Assessment Checklist—Living Systems

Start date _____
End date _____

Student names	Investigation 1, Part 4		
	Planning and carrying out investigations	Analyzing and interpreting data	Systems and system models

Performance Assessment Checklist—Living Systems

Start date _____

End date _____

Investigation 3, Part 3

Student names	Obtaining, evaluating, and communicating information	Systems and system models	Planning and carrying out investigations	Analyzing and interpreting data	Patterns	Scale, proportion, and quantity	Constructing explanations

Assessment Record—Survey/Posttest

Date _____

Student names	1	2	3	4	5	6	7	8	9ab	9cd	10	11	12	13	14	15

FOSS Next Generation
© The Regents of the University of California
Can be duplicated for classroom or workshop use.

Living Systems Module
Assessment Record
No. 4—Assessment Master

Assessment Record—Investigation 1 I-Check

Date _____

Student names	1	2	3	4	5	6ab	6cde	7a	7b	8	9

Assessment Record—Investigation 2 I-Check

Date _____

Student names	1	2	3	4	5	6a	6b	7a	7b	8a	8b

Assessment Record—Investigation 3 I-Check

Date _____

Student names	1a	1b	2	3	4	5	6	7	8	9a	9b	10a	10b

SURVEY/POSTTEST
LIVING SYSTEMS

Name _____

Date _____

1. Study the food web.

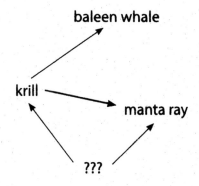

Which organism best replaces the ??? in the food web shown above?

(Mark the one best answer.)

○ **A** A producer

○ **B** A consumer

○ **C** A second-level consumer

○ **D** A decomposer

2. The rabbit population in a certain ecosystem sometimes decreases dramatically. One possible explanation for this decrease is _____.

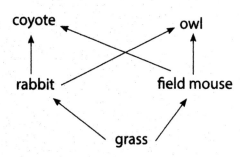

(Mark the one best answer.)

○ **A** an increase in the field mouse population

○ **B** a decrease in the owl population

○ **C** an increase in the grass population

○ **D** an increase in the coyote population

SURVEY/POSTTEST
LIVING SYSTEMS

Name _____

3. Mark **X** next to the statements that describe how marine and terrestrial ecosystems are alike.

   _____ Both rely on the Sun as a source of energy.

   _____ Both rely on plants as the main producers.

   _____ Both include decomposers that breakdown dead material.

   _____ Both include consumers that eat plants and other animals.

4. What is the role played by redworms in an ecosystem?
   Mark **X** next to each sentence that is true.

   _____ They break down organic litter into tiny pieces.

   _____ They are the decomposers that reduce materials to simple minerals.

   _____ They help make room for air and water in soil.

   _____ Their main role is bait for fishing.

5. Which system found in animals is similar to the vascular system in plants?

   (Mark the one best answer.)

   ○ **A** Circulatory system

   ○ **B** Digestive system

   ○ **C** Respiratory system

   ○ **D** Excretory system

SURVEY/POSTTEST
LIVING SYSTEMS

6. Many organisms have features that attract the attention of other organisms.

   Mark **X** next to each phrase that describes why these features are important.

   _____ Attract consumers

   _____ Aid plant pollination

   _____ Aid seed dispersal

   _____ Warn of a bad taste

   _____ Aid tracking at night (especially bright colors)

   _____ Attract a mate

   _____ Distract predators

   _____ Warn when danger is near

7. Match the structures to their functions in the body.

   (Write a letter in front of each structure in the left-hand column.)

_____	Heart	**A**	To reach every cell in the body
_____	Kidney	**B**	To pass nutrients from digested food into the blood
_____	Lung	**C**	To filter waste products from the blood
_____	Stomach	**D**	To carry blood from the heart to the body
_____	Capillary	**E**	To exchange gases with the atmosphere
_____	Artery	**F**	To carry oxygen, food, and nutrients through vessels
_____	Blood	**G**	To pump blood throughout the body
_____	Small intestine	**H**	To break down food with digestive juices and muscles

SURVEY/POSTTEST
LIVING SYSTEMS

Name _____

8. Mark **X** next to each function of the central nervous system.

   _____ Gathering information from the environment

   _____ Breaking food down into small pieces so it is easier to absorb

   _____ Contracting to make arms and legs move

   _____ Coordinating action from the brain to other parts of the body

9. Two students were sitting on a park bench at dusk, eating snacks. A mosquito landed on student A's arm, and she slapped it. When she did that, she spilled a drink into student B's lap, causing him to jump up.

   a. What was the stimulus for student A?

   (Mark the one best answer.)

   ○ **A** The mosquito landing on an arm

   ○ **B** Jumping up

   ○ **C** The slapping motion

   ○ **D** The drink spilling

   b. What was the stimulus for student B?

   (Mark the one best answer.)

   ○ **F** The mosquito landing on an arm

   ○ **G** Jumping up

   ○ **H** The slapping motion

   ○ **J** The drink spilling

   c. What was the response for student A?

   (Mark the one best answer.)

   ○ **A** The mosquito landing on an arm

   ○ **B** Jumping up

   ○ **C** The slapping motion

   ○ **D** The drink spilling

   d. What was the response for student B?

   (Mark the one best answer.)

   ○ **F** The mosquito landing on an arm

   ○ **G** Jumping up

   ○ **H** The slapping motion

   ○ **J** The drink spilling

SURVEY/POSTTEST
LIVING SYSTEMS

Name _____

10. Mark **X** next to the two words that correctly complete this sentence.

    "Plants get the materials they need for growth mainly from _____ and _____."

    _____ Light   _____ Nitrogen

    _____ Soil   _____ Oxygen

    _____ Water   _____ Carbon dioxide

11. What can communities do to improve the environment and help monarch butterflies survive?

    (Mark the one best answer.)

    ○ **A**   Put bird baths in back yards to provide more water for the monarchs.

    ○ **B**   Make sure all grasses are cut short in the summer.

    ○ **C**   Use pesticides to kill as many other insects as possible.

    ○ **D**   Grow plants that butterflies can use for food.

12. Write **1** next to each of the behaviors below that is a result of *instinct*. Write **2** next to behaviors that are *learned*. Write **3** next to behaviors that are *reflexes*.

    (Remember: 1 = instinct; 2 = learned; 3 = reflex)

    _____ Riding a skateboard   _____ Using tools to gather food

    _____ Bears hibernating in winter   _____ Blinking when something comes toward you

    _____ Using a straw to drink   _____ Squirrels gathering acorns

SURVEY/POSTTEST
LIVING SYSTEMS

13. Study the tables below that show the results of a stimulus/response investigation. The object of the investigation is to see how quickly students can move their hands out of the way of a falling cup.

Height of Drop	20 cm	
Stimulus	Vision	
Response	Right hand	
5	x	
4	x	
3	x	x
2	x	x
1	x	x
	Hit	Miss

Height of Drop	20 cm	
Stimulus	Vision	
Response	Left hand	
5		x
4		x
3		x
2	x	x
1	x	x
	Hit	Miss

Is this student right-handed or left-handed?

(Mark the one best answer.)

○ A Right-handed

○ B Left-handed

○ C Does equally well with both hands

○ D There is not enough information here to answer the question.

SURVEY/POSTTEST
LIVING SYSTEMS

Name _____

OPEN-RESPONSE QUESTION

14. Look at the list of organisms in the box.

    - Draw a food web to show how they interact. (In this ecosystem, ants and gophers both eat grass; lizards eat ants, and hawks eat lizards and gophers.)

    - Label each of the organisms in your food web to tell if they are producers, consumers, or decomposers.

Organisms to include in the food web
Ant
Gopher
Grass
Lizard
Bacteria
Hawk

Name _____

SURVEY/POSTTEST
LIVING SYSTEMS

OPEN-RESPONSE QUESTION

15. Two students were talking. Student A was examining the lettuce leaves in a salad he was eating. He said, "Did you know that I am eating energy from the Sun?" Student B was eating a hamburger. She said, "Did you know that I am eating energy that came from the Sun, too?"

    Do you agree with these two students? Explain in words or with a drawing why you agree or disagree with each student.

INVESTIGATION 1 I-CHECK
LIVING SYSTEMS

Name _____

Date _____

1. Study this food web model that shows how organisms interact.

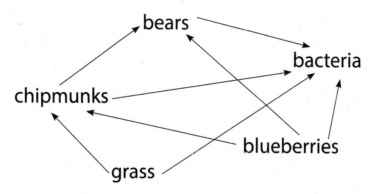

Write **P** in front of the organisms listed below that are *producers*. Write **C** in front of the organisms that are *consumers*. Write **D** in front of the organisms that are *decomposers*.

_____ Bacteria   _____ Chipmunks   _____ Blueberries

_____ Bears   _____ Grass

2. A community of organisms interacting with one another and with nonliving factors in an area is called _____.

   (Mark the one best answer.)

   ○ **A** an environment

   ○ **B** a population

   ○ **C** an ecosystem

   ○ **D** a habitat

3. • An organism that breaks down dead plant and animal material into simple chemicals is a _____.

   • _____ are organisms that make their own food.

   • Organisms that eat other organisms are called _____.

INVESTIGATION 1 I-CHECK
LIVING SYSTEMS

4. Write **M** next to the words that describe matter in an ecosystem. Write **E** next to the words that describe energy in an ecosystem. Write **ME** next to words that describe both matter and energy in an ecosystem.

   _____ Air

   _____ Sunlight

   _____ Food

   _____ Water

5. Tarantulas live in a desert ecosystem. These large spiders rest in their burrows all day. At night they come out to look for food, water, and mates. When day approaches, the tarantulas return to their burrows. What nonliving factors do you think affect the tarantulas' behavior? Mark **X** next to each of those nonliving factors.

   _____ Predators   _____ Temperature

   _____ Wind   _____ The Moon

   _____ Cloud cover   _____ Food supply

   _____ Light   _____ Season

6. Write **T** if the sentence is true; write **F** if the sentence is false.

   _____ a. When producers die, they break down decomposers.

   _____ b. Consumers eat producers or other consumers.

   _____ c. Carnivores eat both plants and animals.

   _____ d. Omnivores eat only insects.

   _____ e. Herbivores eat only plants.

INVESTIGATION 1 I-CHECK
LIVING SYSTEMS

7. Study the food web model shown here.

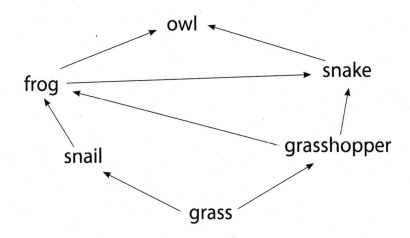

a. What type of organism is missing from this food web?

   (Mark the one best answer.)

   ○ **A**   Producers

   ○ **B**   Consumers

   ○ **C**   Decomposers

   ○ **D**   Carnivores

b. If the frog population suddenly increased, which organism would be most likely to decrease right away?

   (Mark the one best answer.)

   ○ **F**   Owls

   ○ **G**   Snakes

   ○ **H**   Grass

   ○ **J**   Snails

INVESTIGATION 1 I-CHECK
LIVING SYSTEMS

OPEN-RESPONSE QUESTION

8. In most ecosystems, the energy used by all the organisms comes from the Sun. Explain how energy gets from the Sun to a second-level consumer.

INVESTIGATION 1 I-CHECK
LIVING SYSTEMS

Name _____

OPEN-RESPONSE QUESTION

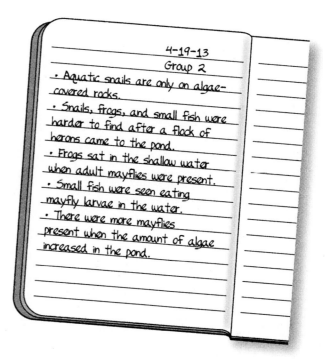

9. In a pond behind the school, students observed several interactions. Read the notebook entries to find out what they observed.

   Choose at least three organisms to construct a food chain. Make sure that you include a producer.

   Construct a food web based on all the organisms observed. Label as specifically as you can which are producers, consumers, and decomposers.

INVESTIGATION 2 I-CHECK
LIVING SYSTEMS

Name _____

Date _____

1. Mark **X** next to each of the components that must be present for plants to make their own food.

    _____ Carbon dioxide   _____ Water

    _____ Oxygen   _____ Light

    _____ Nitrogen   _____ Sugar

2. Mark **X** next to each *product* of photosynthesis.

    _____ Carbon dioxide   _____ Water

    _____ Oxygen   _____ Light

    _____ Nitrogen   _____ Sugar

3. Plants produce food in their _____.

    (Mark the one best answer.)

    ○ **A** roots

    ○ **B** stems

    ○ **C** leaves

    ○ **D** flowers

INVESTIGATION 2 I-CHECK
LIVING SYSTEMS

Name _____

4. Write the names of the parts of the digestive system shown below.

   Word bank: colon, esophagus, large intestine, rectum, small intestine, stomach

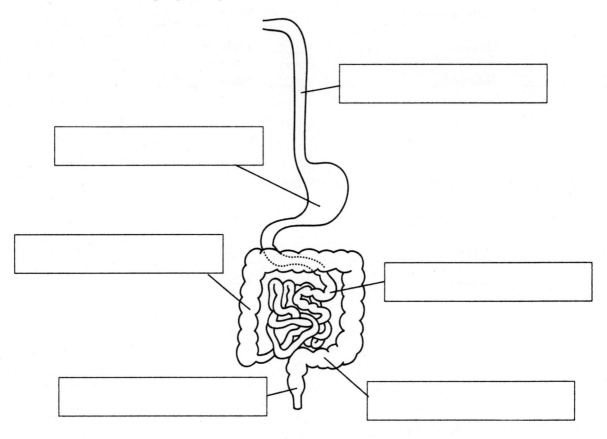

5. Food is anything that people eat that provides _____.

   (Mark the one best answer.)

   ○ **A** energy

   ○ **B** carbon dioxide

   ○ **C** oxygen

   ○ **D** water

INVESTIGATION 2 I-CHECK
LIVING SYSTEMS

6. Two students set up an investigation to see what plants need to produce their own food. They planted a bean in several sets of identical cups of soil. Then they placed them in identical chambers. Chamber A was the control, all factors were present. The other chambers excluded one of the factors.

	Water	Light	Oxygen	Carbon dioxide	Nitrogen
Chamber A	yes	yes	yes	yes	yes
Chamber B	yes	yes	yes	yes	no
Chamber C	yes	yes	yes	no	yes
Chamber D	yes	yes	no	yes	yes
Chamber E	yes	no	yes	yes	yes
Chamber F	no	yes	yes	yes	yes

The students weighed the beans and soil before starting the investigation. After 3 weeks, the students weighed the beans, seedlings (if they were growing), and soil again. Their data are shown here.

	Bean starting mass (g)	Bean ending mass (g)	Soil starting mass (g)	Soil ending mass (g)
Chamber A	500	551	10,000	10,000
Chamber B	500	552	10,000	10,000
Chamber C	500	500	10,000	10,000
Chamber D	500	549	10,000	10,000
Chamber E	500	500	10,000	10,000
Chamber F	500	500	10,000	10,000

a. Mark **X** next to each environmental factor that these data show is needed for plants to make food and increase their mass.

_____ Water _____ Carbon dioxide

_____ Light _____ Nitrogen

_____ Oxygen

b. Mark **X** next to the factor that is the best supporting evidence for your conclusion.

_____ Bean starting mass

_____ Bean ending mass

_____ Soil starting mass

_____ Soil ending mass

INVESTIGATION 2 I-CHECK
LIVING SYSTEMS

OPEN-RESPONSE QUESTION

7. a. What information does a food pyramid describe that a food web might not?

    _____
    _____
    _____
    _____
    _____
    _____
    _____

    b. What level consumer are you? State a claim and provide evidence.

    _____
    _____
    _____
    _____
    _____
    _____
    _____
    _____

INVESTIGATION 2 I-CHECK
LIVING SYSTEMS

Name _____

OPEN-RESPONSE QUESTION

8. A girl's brother told her she should not eat so many cookies because they are made of pure sugar. He suggested she eat a granola bar instead.

   The girl decided to investigate the amount of sugar in Fruity Cream cookies and Cinnamon granola bars. She performed the sugar test on 4 grams of pure sugar, 4 grams of Fruity Cream cookies, and 4 grams of Cinnamon granola bars. The results are pictured below.

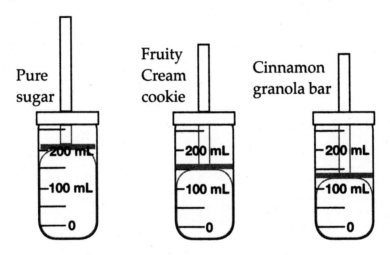

   a. What volume of gas is produced by 4 grams of sugar? _____
      Are Fruity Cream cookies pure sugar? _____
      What is your evidence?

      _____
      _____
      _____
      _____

   b. Do Cinnamon granola bars contain sugar? _____
      How does their sugar content compare to Fruity Cream cookies?

      _____
      _____
      _____
      _____

INVESTIGATION 3 I-CHECK
LIVING SYSTEMS

Name _____

Date _____

1. A student has a collection of leaves from his backyard.

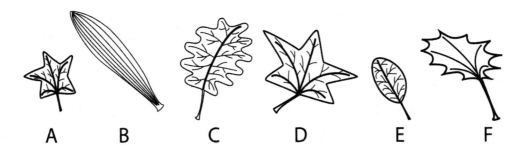

    a.   Circle the letters for the leaves that have pinnate veins.   A B C D E F

          Circle the letters for the leaves that have palmate veins.   A B C D E F

          Circle the letters for the leaves that have parallel veins.   A B C D E F

    b.   Circle the letters for the two leaves that most likely came from the same tree.

        A B C D E F

2. Where do the products of photosynthesis go?

(Write letters next to each component. You may write more than one letter if needed.)

_____ Carbon dioxide

_____ Oxygen

_____ Nitrogen

_____ Water

_____ Light

_____ Sugar

Answer Key

A = Into the air

P = Into the phloem

X = Into the xylem

G = Into the ground

N = Not a product

INVESTIGATION 3 I-CHECK
LIVING SYSTEMS

3. When you look at a cross section of a celery stalk that has been sitting in a vial of red water, only certain parts turn red.

   Why is that part of the celery stalk red?

   These black dots represent where you see the red coloring.

   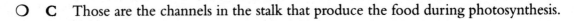

   (Mark the one best answer.)

   ○ A Those are the phloem tubes that carry nutrients to plant cells.

   ○ B Those are the xylem tubes that carry water during transpiration.

   ○ C Those are the channels in the stalk that produce the food during photosynthesis.

   ○ D Those are the veins in the stalk that store the water until it is needed.

4. Why do maple trees need sap?

   (Mark the one best answer.)

   ○ A Maple trees don't need sap. People like it because it is sweet.

   ○ B Sap stores water for maple trees to use when there isn't much rain.

   ○ C Sap provides food to maple tree cells that don't make their own food.

   ○ D Sap helps move water through the maple tree from root to treetop.

INVESTIGATION 3 I-CHECK
LIVING SYSTEMS

5. Write the names of the parts of the human circulatory system shown below.

   Word bank: artery, capillaries, heart, lung, vein

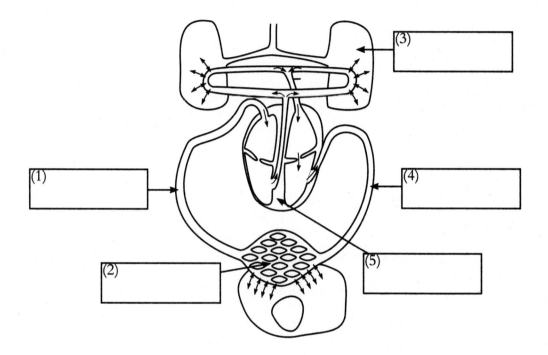

INVESTIGATION 3 I-CHECK
LIVING SYSTEMS

6. What delivers the blood that each human cell needs to survive?

   (Mark the one best answer.)

   ○ **A** Phloem tubes

   ○ **B** Alveoli

   ○ **C** Capillaries

   ○ **D** Arteries

7. Mark **X** next to each phrase that describes why is it important for blood to circulate through the lungs.

   _____ To remove carbon dioxide from the blood

   _____ To remove nitrogen from the blood

   _____ To add water to the blood

   _____ To add oxygen to the blood

   _____ To remove nutrients from the blood

8. What part of the circulatory system carries oxygen to the cells?

   (Mark the one best answer.)

   ○ **A** Veins

   ○ **B** Red blood cells

   ○ **C** White blood cells

   ○ **D** Capillaries

INVESTIGATION 3 I-CHECK
LIVING SYSTEMS

Name _____

OPEN-RESPONSE QUESTION

9. A group of students set up a celery experiment. They recorded the starting mass of two pieces of celery and the volume of water they put in three vials. They put a stalk of celery into two of the vials and used the third vial with water only as a control. The next day, the students recorded the mass of the celery and volume of water in each vial.

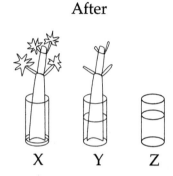

	Vials	Starting volume of water (mL)	Ending volume of water (mL)	Starting mass of celery (g)	Ending mass of celery (g)
X	celery with leaves	25	5	28	29
Y	celery without leaves	25	20	26	27
Z	vial of water	25	24	–	–

a. How much water was lost due to evaporation? _____

How did you figure out how much water was lost due to evaporation?

_____

_____

b. Why was more water lost from vial X than from vial Y?

_____

_____

_____

_____

_____

INVESTIGATION 3 I-CHECK
LIVING SYSTEMS

OPEN-RESPONSE QUESTION

10. Think about two transport systems: the circulatory system in animals and the vascular system in plants.

    a. How are the two systems similar to each other?

    b. How are the two systems different from each other?